Model Discrimination for Nonlinear Regression Models

STATISTICS: Textbooks and Monographs

A Series Edited by

- Vol. 1: The Generalized Jackknife Statistic, *H. L. Gray and W. R. Schucany*
- Vol. 2: Multivariate Analysis, *Anant M. Kshirsagar*
- Vol. 3: Statistics and Society, *Walter T. Federer*
- Vol. 4: Multivariate Analysis: A Selected and Abstracted Bibliography, 1957-1972, *Kocherlakota Subrahmaniam and Kathleen Subrahmaniam* (out of print)
- Vol. 5: Design of Experiments: A Realistic Approach, *Virgil L. Anderson and Robert A. McLean*
- Vol. 6: Statistical and Mathematical Aspects of Pollution Problems, *John W. Pratt*
- Vol. 7: Introduction to Probability and Statistics (in two parts), Part I: Probability; Part II: Statistics, *Narayan C. Giri*
- Vol. 8: Statistical Theory of the Analysis of Experimental Designs, *J. Ogawa*
- Vol. 9: Statistical Techniques in Simulation (in two parts), *Jack P. C. Kleijnen*
- Vol. 10: Data Quality Control and Editing, *Joseph I. Naus* (out of print)
- Vol. 11: Cost of Living Index Numbers: Practice, Precision, and Theory, *Kali S. Banerjee*
- Vol. 12: Weighing Designs: For Chemistry, Medicine, Economics, Operations Research, Statistics, *Kali S. Banerjee*
- Vol. 13: The Search for Oil: Some Statistical Methods and Techniques, *edited by D. B. Owen*
- Vol. 14: Sample Size Choice: Charts for Experiments with Linear Models, *Robert E. Odeh and Martin Fox*
- Vol. 15: Statistical Methods for Engineers and Scientists, *Robert M. Bethea, Benjamin S. Duran, and Thomas L. Boullion*
- Vol. 16: Statistical Quality Control Methods, *Irving W. Burr*
- Vol. 17: On the History of Statistics and Probability, *edited by D. B. Owen*
- Vol. 18: Econometrics, *Peter Schmidt*
- Vol. 19: Sufficient Statistics: Selected Contributions, *Vasant S. Huzurbazar (edited by Anant M. Kshirsagar)*
- Vol. 20: Handbook of Statistical Distributions, *Jagdish K. Patel, C. H. Kapadia, and D. B. Owen*
- Vol. 21: Case Studies in Sample Design, *A. C. Rosander*
- Vol. 22: Pocket Book of Statistical Tables, *compiled by R. E. Odeh, D. B. Owen, Z. W. Birnbaum, and L. Fisher*
- Vol. 23: The Information in Contingency Tables, *D. V. Gokhale and Solomon Kullback*

Vol. 24: Statistical Analysis of Reliability and Life-Testing Models: Theory and Methods, *Lee J. Bain*
Vol. 25: Elementary Statistical Quality Control, *Irving W. Burr*
Vol. 26: An Introduction to Probability and Statistics Using BASIC, *Richard A. Groeneveld*
Vol. 27: Basic Applied Statistics, *B. L. Raktoe and J. J. Hubert*
Vol. 28: A Primer in Probability, *Kathleen Subrahmaniam*
Vol. 29: Random Processes: A First Look, *R. Syski*
Vol. 30: Regression Methods: A Tool for Data Analysis, *Rudolf J. Freund and Paul D. Minton*
Vol. 31: Randomization Tests, *Eugene S. Edgington*
Vol. 32: Tables for Normal Tolerance Limits, Sampling Plans, and Screening, *Robert E. Odeh and D. B. Owen*
Vol. 33: Statistical Computing, *William J. Kennedy, Jr. and James E. Gentle*
Vol. 34: Regression Analysis and Its Application: A Data-Oriented Approach, *Richard F. Gunst and Robert L. Mason*
Vol. 35: Scientific Strategies to Save Your Life, *I. D. J. Bross*
Vol. 36: Statistics in the Pharmaceutical Industry, *edited by C. Ralph Buncher and Jia-Yeong Tsay*
Vol. 37: Sampling from a Finite Population, *J. Hajek*
Vol. 38: Statistical Modeling Techniques, *S. S. Shapiro*
Vol. 39: Statistical Theory and Inference in Research, *T. A. Bancroft and C.-P. Han*
Vol. 40: Handbook of the Normal Distribution, *Jagdish K. Patel and Campbell B. Read*
Vol. 41: Recent Advances in Regression Methods, *Hrishikesh D. Vinod and Aman Ullah*
Vol. 42: Acceptance Sampling in Quality Control, *Edward G. Schilling*
Vol. 43: The Randomized Clinical Trial and Therapeutic Decisions, *edited by Niels Tygstrup, John M. Lachin, and Erik Juhl*
Vol. 44: Regression Analysis of Survival Data in Cancer Chemotherapy, *Walter H. Carter, Jr., Galen L. Wampler, and Donald M. Stablein*
Vol. 45: A Course in Linear Models, *Anant M. Kshirsagar*
Vol. 46: Clinical Trials: Issues and Approaches, *edited by Stanley H. Shapiro and Thomas H. Louis*
Vol. 47: Statistical Analysis of DNA Sequence Data, *edited by B. S. Weir*
Vol. 48: Nonlinear Regression Modeling: A Unified Practical Approach, *David A. Ratkowsky*
Vol. 49: Attribute Sampling Plans, Tables of Tests and Confidence Limits for Proportions, *Robert E. Odeh and D. B. Owen*
Vol. 50: Experimental Design, Statistical Models, and Genetic Statistics, *edited by Klaus Hinkelmann*
Vol. 51: Statistical Methods for Cancer Studies, *edited by Richard G. Cornell*
Vol. 52: Practical Statistical Sampling for Auditors, *Arthur J. Wilburn*
Vol. 53: Statistical Signal Processing, *edited by Edward J. Wegman and James G. Smith*
Vol. 54: Self-Organizing Methods in Modeling: GMDH Type Algorithms, *edited by Stanley J. Farlow*
Vol. 55: Applied Factorial and Fractional Designs, *Robert A. McLean and Virgil L. Anderson*
Vol. 56: Design of Experiments: Ranking and Selection, *edited by Thomas J. Santner and Ajit C. Tamhane*
Vol. 57: Statistical Methods for Engineers and Scientists. Second Edition, Revised and Expanded, *Robert M. Bethea, Benjamin S. Duran, and Thomas L. Boullion*
Vol. 58: Ensemble Modeling: Inference from Small-Scale Properties to Large-Scale Systems, *Alan E. Gelfand and Crayton C. Walker*
Vol. 59: Computer Modeling for Business and Industry, *Bruce L. Bowerman and Richard T. O'Connell*
Vol. 60: Bayesian Analysis of Linear Models, *Lyle D. Broemeling*

Vol. 61: Methodological Issues for Health Care Surveys, *Brenda Cox and Steven Cohen*

Vol. 62: Applied Regression Analysis and Experimental Design, *Richard J. Brook and Gregory C. Arnold*

Vol. 63: Statpal: A Statistical Package for Microcomputers – PC-DOS Version for the IBM PC and Compatibles, *Bruce J. Chalmer and David G. Whitmore*

Vol. 64: Statpal: A Statistical Package for Microcomputers – Apple Version for the II, II+, and IIe, *David G. Whitmore and Bruce J. Chalmer*

Vol. 65: Nonparametric Statistical Inference, Second Edition, Revised and Expanded, *Jean Dickinson Gibbons*

Vol. 66: Design and Analysis of Experiments, *Roger G. Petersen*

Vol. 67: Statistical Methods for Pharmaceutical Research Planning, *Sten W. Bergman and John C. Gittins*

Vol. 68: Goodness-of-Fit Techniques, *edited by Ralph B. D'Agostino and Michael A. Stephens*

Vol. 69: Statistical Methods in Discrimination Litigation, *edited by D. H. Kaye and Mikel Aickin*

Vol. 70: Truncated and Censored Samples from Normal Populations, *Helmut Schneider*

Vol. 71: Robust Inference, *M. L. Tiku, W. Y. Tan, and N. Balakrishnan*

Vol. 72: Statistical Image Processing and Graphics, *edited by Edward J. Wegman and Douglas J. DePriest*

Vol. 73: Assignment Methods in Combinatorial Data Analysis, *Lawrence J. Hubert*

Vol. 74: Econometrics and Structural Change, *Lyle D. Broemeling and Hiroki Tsurumi*

Vol. 75: Multivariate Interpretation of Clinical Laboratory Data, *Adelin Albert and Eugene K. Harris*

Vol. 76: Statistical Tools for Simulation Practitioners, *Jack P. C. Kleijnen*

Vol. 77: Randomization Tests, Second Edition, *Eugene S. Edgington*

Vol. 78: A Folio of Distributions: A Collection of Theoretical Quantile-Quantile Plots, *Edward B. Fowlkes*

Vol. 79: Applied Categorical Data Analysis, *Daniel H. Freeman, Jr.*

Vol. 80: Seemingly Unrelated Regression Equations Models : Estimation and Inference, *Virendra K. Srivastava and David E. A. Giles*

Vol. 81: Response Surfaces: Designs and Analyses, *Andre I. Khuri and John A. Cornell*

Vol. 82: Nonlinear Parameter Estimation: An Integrated System in BASIC, *John C. Nash and Mary Walker-Smith*

Vol. 83: Cancer Modeling, *edited by James R. Thompson and Barry W. Brown*

Vol. 84: Mixture Models: Inference and Applications to Clustering, *Geoffrey J. McLachlan and Kaye E. Basford*

Vol. 85: Randomized Response: Theory and Techniques, *Arijit Chaudhuri and Rahul Mukerjee*

Vol. 86: Biopharmaceutical Statistics for Drug Development, *edited by Karl E. Peace*

Vol. 87: Parts per Million Values for Estimating Quality Levels, *Robert E. Odeh and D. B. Owen*

Vol. 88: Lognormal Distributions: Theory and Applications, *edited by Edwin L. Crow and Kunio Shimizu*

Vol. 89: Properties of Estimators for the Gamma Distribution, *K. O. Bowman and L. R. Shenton*

Vol. 90: Spline Smoothing and Nonparametric Regression, *Randall L. Eubank*

Vol. 91: Linear Least Squares Computations, *R. W. Farebrother*

Vol. 92: Exploring Statistics, *Damaraju Raghavarao*

Vol. 93: Applied Time Series Analysis for Business and Economic Forecasting, *Sufi M. Nazem*

Vol. 94: Bayesian Analysis of Time Series and Dynamic Models, *edited by James C. Spall*

Vol. 95: The Inverse Gaussian Distribution: Theory, Methodology, and Applications, *Raj S. Chhikara and J. Leroy Folks*

Vol. 96: Parameter Estimation in Reliability and Life Span Models, *A. Clifford Cohen and Betty Jones Whitten*

Vol. 97: Pooled Cross-Sectional and Time Series Data Analysis, *Terry E. Dielman*

Vol. 98: Random Processes: A First Look, Second Edition, Revised and Expanded, *R. Syski*

Vol. 99: Generalized Poisson Distributions: Properties and Applications, *P.C. Consul*

Vol. 100: Nonlinear L_p-Norm Estimation, *Rene Gonin and Arthur H. Money*

Vol. 101: Model Discrimination for Nonlinear Regression Models, *Dale S. Borowiak*

Vol. 102: Applied Regression Analysis in Econometrics, *Howard E. Doran*

Vol. 103: Statistical Methodology in the Pharmaceutical Sciences, *Donald A. Berry*

ADDITIONAL VOLUMES IN PREPARATION

Model Discrimination for Nonlinear Regression Models

DALE S. BOROWIAK

Department of Mathematical Sciences
The University of Akron
Akron, Ohio

MARCEL DEKKER, INC.   New York and Basel

Library of Congress Cataloging-in-Publication Data

Borowiak, Dale S.
Model discrimination for nonlinear regression models / Dale S. Borowiak.
p. cm. -- (Statistics, textbooks and monographs; vol. 101)
Bibliography: p.
Includes index.
ISBN 0-8247-8053-1 (alk. paper)
1. Regression analysis. I. Title. II. Series: Statistics, textbooks and monographs ; v. 101.
QA278.2.B686 1989
519.5'36--dc19 88-32362
CIP

MARCEL DEKKER, INC.
270 Madison Avenue, New York, New York 10016

Current printing (last digit):
10 9 8 7 6 5 4 3 2 1

PRINTED IN THE UNITED STATES OF AMERICA

For my family,
where each lives for the other
and all live for God.

Preface

It is common in scientific study to construct a set of proposed nonlinear regression models for empirical data. It then becomes the job of the engineer, sociologist, or consulting statistician to choose among these competing models. When these models are based on physical laws and properties, model selection amounts to choosing among contending theories. Often an efficient model in terms of its predictability is the main goal. The aim of this book is to provide the practitioner and statistician with techniques and basic theory in the area of model discrimination.

This text discusses the dual philosophy present in statistical modeling literature. The first concept, one that is common to the social sciences, is that no one model is correct and, therefore, the modeler searches for a parsimonious model. Diagnostic statistics, referred to as discrimination functions, are constructed and used in the discrimination process. These measure a proposed model's two basic properties, namely, accuracy and, in the case of fitted models, stability. The second concept, more common in the engineering sciences, involves direct comparisons of the proposed models. Out of the set of prospective models, the true model is assumed to exist. Here, discrimination methods based on correct and incorrect selection probabilities are constructed.

The first chapter introduces the basic ideas of model discrimination. Under data normality assumptions, admissible models and selection probabilities are discussed. The situation of completely determined models is considered in the second chapter. Both simultaneous and sequential model selection methods are explored. The fitted or estimated model situation is the topic of the third chapter. Diagnostic statistics, as well as formal hypothesis test procedures, are presented. When the standard nonlinear modeling assumptions, such as data normality, are violated, modern resampling methods can be utilized. This is explored in the fourth chapter, where the jackknife, bootstrap, and model simulation techniques are applied to the previously proposed model discrimination procedures.

The reader is assumed to be familiar with basic statistical theory, such as estimation, distributional theory, and hypothesis testing. For the research statistician, proofs of the various theory and properties presented are included. The guiding principle, however, is that of usefulness to the practitioner interested in model discrimination. For this reason, the many model selection methods, along with examples, can be easily extracted from the text without hindrance from the proofs.

I would like to take this opportunity to thank Marcel Dekker, Inc., for showing interest in my work, thus enabling this text to be written. In particular, I am indebted to Vickie Kearn and Barbara Towar for their assistance.

Dale S. Borowiak

Contents

Model Discrimination for Nonlinear Regression Models

1

Introduction

1.1 MODEL DISCRIMINATION

In analyzing physical systems or examining variable relationships, the researcher is aided by the construction of one or more competing models. Underlying each model may be a particular theory or set of theories. The engineer may have several types of diffusion or compartmental models in mind, whereas the social scientist may consider contrasting correlation models in connection with path analysis. The selection of a best model from the collection of constructed models is often a choice among competing theories and is based on empirical results obtained from sample data. We refer to the theory of selecting among rival models based on sample information as model discrimination.

Two main properties characterize a proposed model's usefulness in data analysis. First is the model's fit to empirical data. Second, in the case of models which contain unknown parameters to be estimated, is the dependence of the fitted model, or estimated parameters, on the particular observed data set. This second property we call model stability. In constructing model selection procedures and tests, these two modeling traits must be taken into account.

Based on the fit of rival models, Atkinson (1969) suggested three hypothetical points of view for statistical testing in model discrimination. These are summarized by the following questions they attempt to answer:

A1. Under the assumption that one particular model is the true one, is there evidence that one of the other models fits the data better?

A2. Under the assumption that the models fit the data equally well, is there statistical evidence to the contrary?

A3. Assuming that one of the models is the true one, is there sufficient evidence to make a selection?

The merit of these basic questions holds when a model's stability and its fit are evaluated. Throughout this text we often refer to these basic model discrimination starting points, which add structure to the entire investigation, as A1, A2, and A3.

In specific settings the application of these questions leads to the construction of model assessment statistics and model selection procedures. Under A1 or A3 a hypothesized model is assumed to be correct, and formal test procedures directly contrasting models yielding correct and incorrect selection probabilities are constructed. Since no model is assumed to be correct under

A2, less structure exists and statistics which assess the models are constructed. The goal of this text is to explore the many resulting discrimination methods.

1.2 REGRESSION MODELS

In data modeling a response variable Y is related to a set of predictor variables x_r, $r = 1, \ldots, p$, which are considered fixed through some functional form, denoted f. If the predictor variables are random variables, commonly called covariates, they are denoted X_r, for $1 \leq r \leq p$. The structure of f often is determined by the mechanisms and theories involved in the physical process or system at hand. For a specified set of x_r, $1 \leq r \leq p$, a proposed model, if correct, assumes the mean of Y, that is, $E(Y) = f(\mathbf{x})$, where $\mathbf{x} = (x_1, \ldots, x_r)$. This construction is referred to as modeling for mean structure.

In general model discrimination problems the researcher is faced with k competing models, denoted by f_i for $i = 1, \ldots, k$. The set of rival models is defined by $\mathcal{S}$. Any model f_i from $\mathcal{S}$ is associated with a function of $\mathbf{x}$ which (when $\mathbf{x}$ is suppressed) we denote by $f(i)$. The model function $f(i)$ may or may not contain parameters to be estimated from the observed data. If no "unknown" parameters exist, the model is referred to as fixed or completely determined. The presence of unspecified parameters requires that the model be fitted to the data.

Samples at various combinations or locations of $\mathbf{x}$, denoted $\mathbf{x}_j = (x_{jl}, \ldots, x_{jp})$, are taken, and associated responses Y_j are measured for $j = l, \ldots, n$. For model f_i, from $\mathcal{S}$, mean structure modeling results in the regression model

$$Y_j = f_j(i) + e_j(i) \tag{1.2.1}$$

where $f_j(i)$ is the value of $f(i)$ evaluated at $\mathbf{x}_j$ and $e_j(i)$ is the associated error term such that $E[e_j(i)] = 0$ for all $j = 1, \ldots, n$. Because of the nature of (1.2.1), the predictor variables are often called independent variables while the response variable is commonly referred to as the dependent variable. The model function $f_j(i)$ is referred to as the deterministic part of (1.2.1). Also, as before, the components of $\mathbf{x}_j$ are considered as fixed quantities.

Models of the form (1.2.1) are often nonlinear in unknown parameters. This topic is taken up in detail in Chapter 3.

Example 1.2.1

For model f_i consider a model function based on r_i independent variables. The model function is linear, in parameters, when (1.2.1) holds and $f_j(i) = \mathbf{x}(i)\boldsymbol{\beta}(i)$ for $n \times (r_i + 1)$ design matrix $\mathbf{x}(i)$, in which the rows comprise a 1 followed by $\mathbf{x}_j$, and $r_i \times 1$ vector of unknown parameters $\boldsymbol{\beta}(i)$. Here $n \geq r_i + 1$. When the model is of full rank, i.e., the rank of $\mathbf{x}(i)$ is $r_i + 1$, it is called a linear regression model. These models have been extensively studied, and we refer to Draper and Smith (1981) for a full treatment.

Although they are less tractable, modeling schemes are not restricted to those with mean structure. In a more general setting we might model for median structure (Problem 1.1). Models of the form (1.2.1) are said to possess errors with additive structure. A multiplicative error-structured model is represented by

$$Y_j = f_j(i)e_j(i) \tag{1.2.2}$$

This model can be transformed to an "additive" model by taking the logarithm of both sides (see Problem 1.2). The resulting model can be judged based on its main characteristics, fit and stability.

Remark 1.2.1

Since the true error structure of a model is unknown, model functions in which both forms, (1.2.1) and (1.2.2), result in equally efficient models are desirable. However, this is generally not the case, and we refer to Ratkowsky (1983, pp. 61-64) for an example.

In this text we restrict our investigation to models, by either design or transformation, of the type (1.2.1). When f_i, from $\mathcal{S}$, is considered, not only is mean structure assumed but, in addition, normality is imposed. Hence, if f_i represents the system being modeled, we assume $e_j(i)$ are i.i.d. $\sim n(0, \sigma^2)$ (independent and identically distributed normal with mean zero and variance σ^2).

Also, when competing models are not directly contrasted, such as under A2, and only one particular model from $\mathcal{S}$ is under discussion, the index i will be dropped; that is, $f_j(i)$ and $e_j(i)$ will be written f_j and e_j. This notation will be employed throughout the text.

1.3 TRUE MODELS

In model discrimination we are faced with the task of selecting among k competing models. If we take up either question A1 or A3, a designation of a true model must be made. How are we to define the model which best reflects the nature of the system being examined? To answer this question we assume modeling scheme (1.2.1) and accompanying assumptions. For a particular model function $f(i)$, $1 \leq i \leq k$, we assume mean structure and normality of the error term.

A problem arises in the situation where models contain unknown parameters. Here, the true model function may not be unique in that some models may be reductions of others. In addition, when models lack theoretical justification a designation of a true model may not be desirable. This is often the case in the social sciences, where subsets of linear regression models may appear to be equivalent in terms of their observed properties. In this case, Atkinson's question A2 may be more appropriate.

Based on mean structure modeling, a model f_i, from $\mathcal{S}$, is said to be admissible for the data set if $E[e_j(i)] = 0$ for all $j = 1, \ldots, n$. In the completely determined model case the uniqueness of the models implies the existence of only one admissible model. This model is defined as the true model. When unknown parameters are present, there may be more than one admissible model. In this case, the true model is defined as the admissible model contained in all other admissible models.

Example 1.3.1

Consider the case of a linear regression model introduced in Example 1.2.1. Including all p possible independent variables ($r_i = p$) leads to the "full" model. By restricting the number of predictor variables we reduce the design matrix by eliminating corresponding columns, thereby producing other linear regression models. Let $\mathcal{S}$ contain all possible 2^p models. A regression model f_i in which all predictor variables are contained in another model f_q is called nested in f_q, denoted $f_i \subset f_q$. Thus, if f_i is admissible, then so is any f_q where $f_i \subset f_q$. Also, if the full model is admissible then $\mathcal{S}$ contains a true model.

If either question A1 or A3 is asked, then we assume $\mathcal{S}$ contains a unique true model. In this case, the ideas of correct and incorrect selections along with their corresponding probabilities may be explored. If question A2 is put forth, then we need not define a true model and, for comparison, model characteristics are contrasted.

1.4 DISCRIMINATION FUNCTIONS

From any of Atkinson's starting questions, A1, A2, or A3, the modeler constructs statistics based on the responses and observed model values to be used in the discrimination procedure. These functions, or statistics, may be used to select best or true models, eliminate unlikely models, or designate when models are too similar to be chosen as true or false. We refer to these functions as discrimination functions, denoted DFs. Under A1 or A3 a discrimination method, referred to as DM, selects a best model from $\mathcal{S}$ and may use one DF or employ a series of DFs. The choice of the DM utilized will depend on the particular circumstances at hand.

Commonly, DFs are based on the observed errors or residuals of the rival models. This is particularly true when the fit of a model is considered. For f_i from $\mathcal{S}$, based on (1.2.1), the residuals are defined by

$$R_j(i) = Y_j - f_j(i) \tag{1.4.1}$$

for $1 \leq j \leq n$. The vector of residuals is denoted

$$\mathbf{R}(i)' = (R_1(i), \ldots, R_n(i)) \tag{1.4.2}$$

Here $\mathbf{A}'$ is the transpose of matrix $\mathbf{A}$. The realizations, or observed values, of these are $R_j(i) = r_j(i)$ and $\mathbf{R}(i) = \mathbf{r}(i)$. Also, the vector of model function values is given by $\mathbf{f}(i)' = (f_1(i), \ldots, f_n(i))$. We notc that in the case of unknown parameters the model function $f_j(i)$, in (1.4.1), and $\mathbf{f}(i)$ must be estimated. This designation will be made when the "estimated model" setting is discussed in Chapter 3.

A DF which is based on a quadratic function of residuals is called a quadratic discrimination function or QDF. For model f_i

from $\mathcal{S}$, a QDF takes the form

$$Q(i) = \mathbf{R}(i)'\mathbf{S}(i)\mathbf{R}(i) + c(i) \tag{1.4.3}$$

where $\mathbf{S}(i)$ is an $n \times n$ positive semidefinite matrix, $c(i)$ is a constant, and $\mathbf{R}(i)$ is defined by (1.4.2). If we take $\mathbf{S}(i)$ to be the identity matrix $\mathbf{I}$ and define $c(i)$ to be zero, then (1.4.3) becomes the residual sum of squares for f_i, denoted RSS(f_i). The larger the value of $Q(i)$, the more model f_i is suspect (Problem 1.3).

Remark 1.4.1

In the estimated model case (unknown parameters) QDF can be adjusted to equate models with differing numbers of parameters by the choice of $\mathbf{S}(i)$ and $c(i)$. This is especially true for linear regression models (see Problem 1.4).

Consider discrimination from the viewpoint of A1 or A3, where we assume $\mathcal{S}$ contains a true model. The simplest DM is to select, from $\mathcal{S}$, the model with the least QDF. We refer to such a procedure as a least quadratic method or LQM.

Example 1.4.1

Least squares error discrimination is an LQM with $\mathbf{S}(i) = \mathbf{I}$ and $c(i) = 0$ for all $i = 1, \ldots, k$. This DM selects the model with the smallest value of RSS and is referred to as LSE. This method is shown to be adequate for sets of competing linear models with equal numbers of unknown parameters (Problem 1.3).

The main task in the construction of a DM is to define one or more DFs that enable us to carry out tests or procedures in the framework of A1, A2, or A3. Not only are DFs that assess fit, such as (1.4.1), utilized, but also, in the case of estimated models, ones that measure stability are employed.

1.5 SELECTION PROBABILITIES

In this section we discuss the basic structures present when either question A1 or A3 is examined. Here we assume $\mathcal{S}$ contains a true model and a particular DM is employed. This DM selects, with probability one, a unique best model. In this setting, correct and incorrect selections, along with their associated probabilities, are the basic tools in the construction and evaluation of an efficient DM.

There are two decisions inherent in such a DM. If f_i from $\mathcal{S}$ is the true model and it is chosen, a correct selection, CS, occurs, which we denote CS(f_i). On the other hand, if f_i is selected while f_q is the correct model an incorrect selection, IS, is observed, which is then denoted IS($f_i \mid f_q$). Taking all the models of $\mathcal{S}$ into account,

$$\text{IS}(f_i \mid f_q) = \cap\{\text{Selecting } f_i \text{ over } f_m \mid f_q \text{ true}\} \tag{1.5.1}$$

and

$$\text{CS}(f_i) = \cap\{\text{Selecting } f_i \text{ over } f_m \mid f_i \text{ true}\} \tag{1.5.2}$$

where the intersections are over $f_m \in \mathcal{S}$ and $m \neq i$.

When possible, in conjunction with a DM, the choice of the best model is assessed by the probabilities of (1.5.1) and (1.5.2). Under the least favorable model, the probability of an IS, while selecting f_i, is denoted IS(f_i). Hence,

$$P(\text{IS}(f_i)) = \max P(\text{IS}(f_i \mid f_q)) \tag{1.5.3}$$

where the maximum is over $f_q \in S$ and $q \neq i$. Also, the probability of a CS when f_i is chosen is denoted $P(\text{CS}(f_i))$. Probabilities (1.5.3) and $P(\text{CS}(f_i))$ depend on the model selected and are therefore referred to as conditional selection probabilities.

If all possible choices of a hypothesized true model are considered, probabilities that are independent of the model selected can be constructed. When f_q is the true model an IS is the event $\cup \mathrm{IS}(f_i \mid f_q)$, where the union is over $i \neq q$. Thus, the overall probability of an IS is

$$P(\mathrm{IS}) = \max_{1 \leq m \leq k} P\left(\bigcup_{i \neq m} \mathrm{IS}(f_i \mid f_m) \right) \tag{1.5.4}$$

In a like manner, the overall probability of a CS is

$$P(\mathrm{CS}) = \min_{1 \leq i \leq k} P(\mathrm{CS}(f_i)) \tag{1.5.5}$$

The simplest type of DM utilizes DFs that pairwise compare models. In the case of two models, f_i and f_q from $\mathcal{S}$, and a fixed DM, the probability of selecting f_i over f_q under true model f_q is denoted by $P(f_i \mid f_q)$. In the following theorem probability bounds for pairwise procedures are constructed.

Theorem 1.5.1

Assume $\mathcal{S}$ contains a true model. Utilizing a DM which selects a best model with probability one, conditioning on the selected model,

$$P(\mathrm{IS}(f_i)) \leq \max_{q \neq i} P(f_i \mid f_q) \tag{1.5.6}$$

and

$$P(\mathrm{CS}(f_i)) \geq 1 - \sum_{q \neq i} P(f_q \mid f_i) \tag{1.5.7}$$

Further, without conditioning,

$$P(\mathrm{IS}) \leq \max_{1 \leq q \leq k} \sum_{i \neq q} P(f_i \mid f_q) \tag{1.5.8}$$

Proof The proof, which is left as an exercise (Problem 1.5), relies on (1.5.1), (1.5.2), and definitions (1.5.3) and (1.5.4).

In the case of completely determined models, conditional selection probabilities or appropriate bounds can be utilized to test Atkinson's A1 or A3. For flavor we now consider two specific test constructions.

First, consider question A1 where f_i is fixed and competes against all other completely determined models in S. Here we test

$$H_0{:}f_i\text{Correct} \qquad \text{vs.} \qquad H_1{:}f_i\text{Incorrect} \tag{1.5.9}$$

by rejecting H_0 if f_i, $q \neq i$, is selected and, for some predetermined α, $0 \leq \alpha \leq 1$, we find $P(\mathrm{IS}(f_q \mid f_i)) \leq \alpha$. Using (1.5.8), under hypothesized model f_i, we find that the probability of a type one error is bounded above by $\alpha/(k-1)$.

Second, with A3 no particular model is picked for comparisons. Suppose using a DM f_i is selected. We find there is not enough evidence to make a selection if $P(\mathrm{IS}(f_i)) \geq \alpha$ for some α, $0 \leq \alpha \leq 1$. If a selection can be made, then the magnitude of the observed $P(\mathrm{IS}(f_i))$ measures the faith in the selection.

Standard statistical tests which compare rival models can often be put in the framework of A1 or A3. This is particularly true when we are testing for model reduction. The following example illustrates this in an estimated model setting.

Example 1.5.1

Consider comparing nested linear regression models f_i and f_q where $f_i \subset f_q$ (see Example 1.3.1). From viewpoint A1, using general linear model theory, we test (1.5.9) (only two possible models) using the F statistic

$$F = (n - r_q - 1)[\mathrm{SSR}(\hat{f}_i) - \mathrm{SSR}(\hat{f}_q)]/[(r_q - r_i)\,\mathrm{SSR}(\hat{f}_q)] \quad (1.5.10)$$

Here, F has degrees of freedom $r_q - r_i$ and $n - r_q - i$ and we reject H_0:f_i correct when $F \geq F_{(1-\alpha)}$. In this case $P(\mathrm{IS}(f_q)) \leq \alpha$, whereas $P(\mathrm{IS}(f_i))$ is not so easily found.

In the chapters ahead we construct DMs under various conditions. In some cases it is possible to calculate exact values or appropriate bounds for selection probabilities. In this way efficient measures of the DM and resulting selected model arc obtained.

PROBLEMS

1.1 Model f_i possesses median structure if the median of $e_j(i)$ is zero for $j = 1, \ldots, n$.
   (a) What distributions can we put on $e_j(i)$ and assume median structure?
   (b) For a fixed $f_i \in \mathcal{S}$ construct an a priori test of median structure based on the binomial random variable $W = 1$ if $r_j(i) \geq 0$ and 0 if $r_j(i) \leq 0$.

1.2 Transform (1.2.2) into an additive model, (1.2.1), using the natural logarithmic function. What assumptions are on the error term if the transformed model has mean structure and normality?

1.3 Let f_i be a linear regression model with rank $r_i + 1$ (see Example 1.2.1). Suppose $\mathbf{Y} \sim N(\boldsymbol{\mu}(i), \sigma^2\mathbf{I})$. Here f_i is correct if and only if $\boldsymbol{\mu}(i) = \mathbf{x}(i)\boldsymbol{\beta}(i)$.

(a) Letting the estimated model be $\mathbf{E}(i)\mathbf{Y}$ where $\mathbf{E}(i) = \mathbf{x}(i)[\mathbf{x}(i)'\mathbf{x}(I)]^{-1}\mathbf{x}(i)'$ show $E[\mathbf{R}(i)'\mathbf{S}(i)\mathbf{R}(i)] = \sigma^2 \times \mathrm{Tr}[\mathbf{S}(i)(\mathbf{I} - \mathbf{E}(i))] + \boldsymbol{\mu}(i)'(\mathbf{I} - \mathbf{E}(i)]\mathbf{S}(i)[\mathbf{I} - \mathbf{E}(i)]\boldsymbol{\mu}(i)$.

(b) Show that $E[Q(i)]$ is a minimum when f_i is correct, where $Q(i)$ is defined by (1.4.3).

(c) If $\mathbf{S}(i) = \mathbf{I}$ and $c(i) = 0$, when, assuming model correctness, is $E[Q(i)]$ equal for two models f_i and f_q (note $\mathrm{Tr}[\mathbf{E}(i)] = r_i + 1$)?

(d) Define $Q(i)$, given by (1.4.3), such that under the true model, $E[Q(i)]$ are equal for $i = 1, \ldots, k$.

1.4 Consider the linear regression model setting of Example 1.3.1. For constant a the following DFs have been proposed in the literature:

(a) $\mathrm{AIC}(i) = n \log[\mathrm{RSS}(f_i) + ar_i$ (Akaike, 1973).

(b) $C_p(i) = \mathrm{RSS}(f_i)/\sigma^2 + am - n$ (Mallows, 1973).

(c) $\mathrm{FPE}(i) = [n + a(r_i - 1)]\mathrm{RSS}(f_i)$ (Akaike, 1970).

(d) $\mathrm{PSS}(i) = \mathbf{R}'(i)[\mathbf{I} - \mathbf{M}(i)]^2\mathbf{R}(i)$ where $\mathbf{M}(i)$ is a diagonal matrix consisting of the diagonal elements of $\mathbf{E}(i)$ (Allen, 1971).

(e) $\mathrm{GIC}(i) = n \log[\mathrm{RSS}(f_i)] + a_n r_i$ where $\{a_n\}$ is a sequence of constants such that $\lim_{n\to\infty} a_n = \infty$ and $\lim_{n\to\infty} a_n/n = 0$ (Schwartz, 1978).

Write the above DFs, when possible, in the form of QDFs. Using Problem 1.3, find the expected value of each under the true model assumption. Do they adjust for differing r_i?

1.5 Prove Theorem 1.5.1

1.6 For two deterministic models, f_i and f_q, consider the LQM where $\mathbf{S}(i) = \mathbf{S}$ and $c(i) = 0$ for $i = 1, 2$.

(a) Show f_i is selected over f_q if

$$2\mathbf{f}(q,i)'\mathbf{SR}(i) \leq \mathbf{f}(i,q)'\mathbf{Sf}(i,q) \tag{1.6.1}$$

where $\mathbf{f}(i,q) = \mathbf{f}(i) - \mathbf{f}(q)$.

(b) Since f_i and f_q are deterministic, $\mathbf{R}(i) \sim N(0, \sigma^2 I)$ for true model f_i. Find $P(\text{IS}(f_i))$ and note that it is independent of the model selected.

(c) Taking up A3, construct a test of H_0: Not Enough Evidence to Make a Selection. How do we measure a selection?

1.7 Consider linear regression models f_1, f_2, and f_3 where $f_1 \subset f_2$ and $f_1 \subset f_3$.

(a) Construct a test of (1.5.9) at significance level α where $i = 1$.

(b) If $n = 10, r_1 = 1$, and $r_2 = r_3 = 2$, suppose $\text{RSS}(f_1) = 10$, $\text{RSS}(f_2) = 4$, and $\text{RSS}(f_3) = 2$. Run the test constructed in part (a) at significance level $\alpha = .10$.

2

Completely Determined Models

2.1 INTRODUCTION

In this chapter the rival model functions of $\mathcal{S}$ contain no parameters to be estimated from the data and hence are completely determined. These models arise primarily in an engineering context where it is not uncommon to use parameter values estimated from historical sources. Because of the structure present, we take up question A1 or A3 and construct DMs which often lead to selection probabilities, or appropriate bounds. Some of these DMs are extended to the estimated model case in the chapters to follow.

The k models of $\mathcal{S}$ are assumed to take the form of (1.2.1) and normality of the error term is imposed. Hence, under true model f_i, $e_j(i) \sim$ i.i.d.$n(0,\sigma^2)$ for $j = 1, \ldots, n$. Since the model functions are fixed, the error terms are observed and coincide with the residuals, (1.4.1) and (1.4.2). Therefore, under the true model f_i, $\mathbf{R}(i)$ is distributed multivariate normal with zero mean vector and dispersion matrix $\sigma^2\mathbf{I}$, that is $\mathbf{R}(i) \sim N(0, \sigma^2\mathbf{I})$.

Because of the fixed nature of the rival model functions, model stability is not a factor. Hence, only a model's fit to the empirical data plays a part in model selection. For this reason the DMs in this chapter utilize DFs which assess a model's fit to the responses relative to the fit of its competitors.

The size of k affects the efficiency of any DM (consider bounds (1.5.7) and (1.5.8)). For this reason, the researcher will be aided if highly unlikely models are eliminated from S before a DM is employed. A DF, or statistic, based on the RSS to measure a model's lack of fit can be applied to individual models of S. For true model $f_i \in S$ we note that RSS $(f_i)/\sigma^2 \sim \chi^2$ with n degrees of freedom. Model f_i is eliminated from consideration if the lack-of-fit measure

$$\mathrm{LOF}(f_i) = P[\chi^2_{(n)} > \mathrm{rss}(f_i)/\sigma^2] \tag{2.1.1}$$

is small (less than α, $\alpha = .1, .05$, or $.01$), where rss(f_i) is the observed RSS for model f_i. This test of model adequacy is, in theory, under A2, since models are not directly contrasted but simply assessed (by LOF). All models in S at the time a DM is applied are assumed to be reasonable as measured by LOF.

2.2 TWO-MODEL CASE

The earliest work in model discrimination is the two-model procedure of Hoel (1947). Set S contains $k = 2$ completely determined models and viewpoint A1 is taken. Thus, hypothesis (1.5.9) is put forth and a model, say f_i, is considered true under H_0. Hoel constructed a fitted model f^*, defined by $f_j^* = af_j(1) + (1-a)f_j(2)$, to be tested against f_i. The value of a which minimizes RSS(f^*), namely,

$$\hat{a} = \sum_{j=1}^{n} R_j(2)f_j(1,2) \Big/ \left[\sum_{j=1}^{n} f_j(1,2)^2\right] \qquad (2.2.1)$$

is utilized, where $f_j(1,2) = f_j(1) - f_j(2)$ (the circumflex ˆ indicates estimation based on the observed responses). We note that both f_1 and f_2 are nested in f^*. Hoel tested H_0: f_1 correct against H_1: f^* correct using an F statistic with degrees of freedom 1 and $n-1$, given by

$$F = (n-1)[\text{RSS}(f_1) - \text{RSS}(f^*)]/\text{RSS}(f^*) \qquad (2.2.2)$$

An α level test rejects H_0 if $F > F_{(1-\alpha,1,n-1)}$. The derivation of the distribution of (2.2.2) is outlined in Problem 2.1.

Hoel's method assumes a model preference to be designated under H_0. A more even method, with respect to the proposed models, is considered by taking up question A3. Here, either f_1 or f_2 is assumed to be correct and we search for the most efficient DM. To this end, let C_1 be the class of LQM where $\mathbf{S}(i) = \mathbf{S}$ and $c(i) = 0$ for $i = 1, 2$. Further, let C_2 be the class of LQM in C_1 where $\mathbf{S}$ has rank one, that is, $\mathbf{S} = \mathbf{AA}'$, where $\mathbf{A}$ is $n \times 1$. The connection between DM in C_1 and C_2 is explored in the following theorem.

Theorem 2.2.1

Let $k = 2$ for models of the form (1.2.2) where normality of the error term is assumed. Then $C_1 = C_2$ and employing any LQM in

C_1,

$$P(f_1 \mid f_2) = P\{Z > |\mathbf{A}'\mathbf{f}(1,2)|/[(\mathbf{A}'\mathbf{A})^{1/2}2\sigma]\} \tag{2.2.3}$$

where $Z \sim n(0,1)$ and $\mathbf{A} = \mathbf{Sf}(2,1)$.

Proof Clearly $C_2 \subseteq C_1$. To show $C_1 \subseteq C_2$ let $\mathbf{S}$ be positive semidefinite and define $\mathbf{A} = \mathbf{Sf}(2,1)$. Taking $\mathbf{S} = \mathbf{AA}'$ in (1.6.1) yields the original inequality, implying $C_1 \subseteq C_2$. If f_2 is the true model, substituting $\mathbf{R}(1) = \mathbf{R}(2) + \mathbf{f}(2,1)$ into (1.6.1) with $\mathbf{S} = \mathbf{AA}'$ gives $2\mathbf{A}'\mathbf{R}(2) < -\mathbf{A}'\mathbf{f}(2,1)$ if $\mathbf{A}'\mathbf{f}(2,1) > 0$ and $2\mathbf{A}'\mathbf{R}(2) > -\mathbf{A}'\mathbf{f}(2,1)$ if $\mathbf{A}'\mathbf{f}(2,1) < 0$. Since under model f_2, $\mathbf{A}'\mathbf{R}(2) \sim n(0, \mathbf{A}'\mathbf{A}\sigma^2)$, standardization yields (2.2.3).

From Theorem 2.2.1 we note that $P[\mathrm{IS}(f_i)]$ is independent of the model selected. Therefore, employing any LQM from C_1,

$$P(\mathrm{IS}) = P(f_1 \mid f_2) \tag{2.2.4}$$

given by (2.2.3) where $\mathbf{A} = \mathbf{Sf}(2,1)$. The most efficient LQM, in terms of choices for $\mathbf{S}$ through $\mathbf{A}$, is now presented.

Theorem 2.2.2

Under the assumptions of Theorem 2.2.1, the LQM in C_1 which minimizes (2.2.4) is the LSE procedure. With the LSE DM

$$P(\mathrm{IS}) = P\{Z > [\mathbf{f}(1,2)'\mathbf{f}(1,2)]^{1/2}/(2\sigma)\} \tag{2.2.5}$$

Proof The minimum value of (2.2.3) of Theorem 2.2.1 is obtained by maximizing $[\mathbf{A}'\mathbf{f}(2,1)/(\mathbf{A}'\mathbf{A})^{1/2}]^2$. The Cauchy–Schwartz inequality implies $[\mathbf{A}'\mathbf{f}(2,1)]^2/\mathbf{A}'\mathbf{A} \leq \mathbf{f}(2,1)'\mathbf{f}(2,1)$. The proof is completed using (2.2.4) and noting from (1.6.1) that the LQMs with $\mathbf{S} = \mathbf{I}$ and $\mathbf{S} = \mathbf{f}(2,1)\mathbf{f}(2,1)'$ are equivalent.

The optimality of the LSE method can also be demonstrated using the Neyman–Pearson lemma (Problem 2.2).

In the case of known σ^2, the previous theorem allows us to examine question A3. If $P(\text{IS})$ is sufficiently small, we determine that a selection can be made and the confidence in the decision is measured by $P(\text{IS})$.

Example 2.2.1

A physicist compares two models for a free-falling object. The first, f_1, assumes a constant acceleration, whereas the second, f_2, accounts for drag or friction using variable acceleration (Sears, 1944, pp. 59–64). An object is dropped from an initial height of $h(0)$ feet. Applying f_1, the object's height at time t is $h(t) = h(0) + .5gt^2$, where $g = -32$ ft/sec^2 (t in terms of seconds). With f_2, $h(t) = h(0) + .5gt^2 + ct^3/3$, where we take $c = 2$ ft/sec^3. The time to hit the ground is recorded and plays the part of modeling function. Hence, $f(1) = [2h(0)/(-g)]^{1/2}$ and $f(2)$ is calculated numerically using Newton's method (find the root of $h(t)$). The object is then dropped from heights 10, 20, 40, and 60 feet and the time to hit the ground and the model function values are found. These quantities and the RSS are listed in Table 2.1.

The physicist estimates the range of measurements to be .26 sec, and using the six standard deviation rule, i.e., range $= 6\sigma$, $\sigma =$

Table 2.1 Free-Falling Bodies

	Heights, $h(0)$				
Responses	10	20	40	60	RSS
	.85	1.13	1.58	1.97	
f_1	.79	1.12	1.58	1.93	.0053
f_2	.80	1.15	1.64	2.02	.0090

.042. Using (2.1.1), both models are acceptable with LOF(f_1) = .538 and LOF(f_2) = .259. The LSE method is applied and f_1 is chosen. Here $\mathbf{f}(1,2)'\mathbf{f}(1,2) = .0127$, and from (2.2.5) we find $P(\mathrm{IS}) = P(Z > 1.31) = .095$. Hence, from the viewpoint of A3 we reject H_0: Not Enough Evidence to Make a Selection at significance level $\alpha = .10$. The confidence in the chosen model, f_1, is measured by $P(f_1 \mid f_2) < .1$.

When comparing two completely determined models, Theorem 2.2.2 instructs us to use the LSE DM even when σ^2 is unknown. In this case (2.2.5) cannot be calculated. To estimate σ^2 we might turn to methods which assume a true model, but then $P(\mathrm{IS}) = P(f_i \mid f_q)$ is likely to be overestimated when, in fact, the true model is chosen. This is because the hypothesized true model f_q, being incorrect, leads to a larger than expected estimate of the variance (see Problem 2.3). Hence, in estimating σ^2, an estimate not influenced by the residuals of the rival models is needed.

An estimate of the variance, independent of an assumed model, exists if the data set contains "repeat" data points. In this setting more than one response is measured at some of the locations. The data takes the form $(Y_{qr}, \mathbf{x}_{qr})$, where r indexes the repetitions, $r = 1, \ldots, n_q$, at distinct locations $q = 1, \ldots, s$. For a fixed q, the location vectors $\mathbf{x}_{qr}$ are identical (same independent variable values) for $r = 1, \ldots, n_q$. Thus $\Sigma n_q = n$. Here, an unbiased estimate of σ^2 is

$$\hat{\sigma}^2 = \sum_{q=1}^{s} \sum_{r=1}^{n_q} (Y_{qr} - Y_{q.})^2 / m \tag{2.2.6}$$

where $Y_{q.} = \Sigma Y_{q,}/r_q$ and $m = \sum_{q=1}^{s}(r_q - 1)$. In the linear regression model setting (2.2.6) is often called a pure estimate of the variance.

This estimator can be used to adjust the preliminary measure LOF given by (2.1.1) (see (2.6.1)).

In connection with the LSE procedure, substituting $\hat{\sigma} = (\hat{\sigma}^2)^{1/2}$ into (2.2.5) yields an estimate of $P(\mathrm{IS})$ given by

$$\hat{P}(\mathrm{IS}) = P\{Z > [\mathbf{f}(1,2)'\mathbf{f}(1,2)]^{1/2}/(2\hat{\sigma})\} \tag{2.2.7}$$

The expected value of this estimate is examined in the theorem which follows.

Theorem 2.2.3

Under the conditions of Theorem 2.2.1, consider the LSE procedure. Let Bias $= E[\hat{P}(\mathrm{IS}) - P(\mathrm{IS})]$. Then

$$E[\hat{P}(\mathrm{IS})] = \int_0^\infty P[Z > \delta/g(W)]\,dFW \tag{2.2.8}$$

and

$$\begin{aligned}\text{Bias} = &\int_{B1} P(\delta/g(W) \le Z \le \delta/(2\sigma))\,dFW \\ &- \int_{B2} P(\delta/(2\sigma) \le Z \le \delta/g(W))\,dFW\end{aligned} \tag{2.2.9}$$

where in the above Riemann–Stieltjs integrals $g(W) = (4\sigma^2 W/m)^{1/2}$, $W \sim \chi^2_{(m)}$ (W is distributed chi square with m degrees of freedom), $\delta = (\mathbf{f}(1,2)'\mathbf{f}(1,2))^{1/2}$, $B1 = \{W, W/m \ge 1\}$, and $B2 = \{W, W/m < 1\}$.

Proof For $\hat{\sigma}^2$ given by (2.2.6), we note that $W = \hat{\sigma}^2 m/\sigma^2 \sim \chi^2_{(m)}$ and $\hat{\sigma}^2$, and hence W, is independent of $\mathbf{R}(1)$ and $\mathbf{R}(2)$. Taking the expected value, we obtain (2.2.8). Using (2.2.8), we find that (2.2.9) follows.

This theorem indicates that $\hat{P}(\text{IS})$, given by (2.2.7), is a biased estimator. However, the magnitude of the bias may be small. We note that W/m converges in probability to one as m increases, that is, $W/m \xrightarrow{P} 1$ as $m \to \infty$. From (2.2.9) we see that the bias will be small if either m is large or $\delta/(2\sigma)$ is far in the tail of a standard normal random variable.

Example 2.2.2

Two completely determined models are compared using LSE. For various values of $\gamma = \delta/(2\sigma)$ and a range of degrees of freedom m, the bias, given by (2.2.9), is calculated. The results are listed in Table 2.2. We find that $\hat{P}(\text{IS})$ underestimates the true value of $P(\text{IS})$ in terms of its expected value. For the ranges of γ and m listed in Table 2.2, $|\text{Bias}| < .044$. Further, in accordance with (2.2.9), for γ greater than 1.25 or m larger than 4 the amount of bias is minimal (found to be less than .01 in absolute value).

In practice, when (2.2.6) is utilized in the construction of tests of A3, a slight adjustment may be in order. Our estimate of $\hat{P}(\text{IS})$ might be adjusted upward a small amount. Also, the jackknife method of Tukey (1958), as suggested by Quenouille (1949), may be employed to reduce the bias. This procedure is discussed in detail in Chapter 4 and an example is outlined in Problem 2.4.

Remark 2.2.1

In the discriminant analysis setting the QDF given by (1.4.3) has been applied by Bayne and Tan (1981). A vector taking the place of the residual vector is to be classified into one of two multivariate normal populations. In the QDF $\mathbf{S}(i)$ is the inverse of the dispersion matrix of the particular models. Under various distributions, selection probabilities are explored.

In the situation where no repeat data points exist, a conservative estimate of σ may be used. When a set of comparable models, in

Table 2.2 Biasedness of $P(\mathrm{IS}(f))$

γ	m	$P(\mathrm{IS}(f_i))$	Bias
.25	1	.4205	−.0338
.25	2	.4205	−.0308
.25	4	.4205	−.0187
.50	1	.3086	−.0378
.50	2	.3086	−.0340
.50	4	.3086	−.0169
.75	1	.2415	−.0431
.75	2	.2415	−.0439
.75	4	.2415	−.0304
1.00	1	.1587	−.0199
1.00	2	.1587	−.0219
1.00	4	.1587	−.0139
1.25	1	.1146	−.0130
1.25	2	.1146	−.0171
1.25	4	.1146	−.0137

term of RSS, is examined, the maximum estimate based on each underlying model may serve as σ^2. This method is demonstrated in Problem 2.9.

2.3 SIMULTANEOUS METHODS

The case of testing among the k, $k > 2$, completely determined models when all models are compared simultaneously is considered. The first DMs of this type examine question A2, whether the models are statistically different. In these procedures representative models under the two alternatives, sameness and uniqueness, are constructed and then compared. Two such procedures are now examined.

The test given by Williams (1959) is based on a test of homogeneity of variances by Wilks (1946). Let the model constructed by averaging at each location over all proposed models be f. Hence, the model function values are $\bar{f}_j = \sum_{J=1}^{n} f_j(i)/k$. This model is compared to f^* defined by the vector of values $\mathbf{f}^* = \mathbf{x}\boldsymbol{\beta}^*$, where $\mathbf{1}'\boldsymbol{\beta}^* = 1$, $\mathbf{1}' = (1,\ldots,1)$, and $\mathbf{x}$ is the $n \times k$ matrix consisting of rows of $(f_j(1),\ldots,f_j(k)), j = 1,\ldots,n$. The value of vector $\boldsymbol{\beta}^*$ which minimizes RSS, given by

$$\hat{\boldsymbol{\beta}}^* = (\mathbf{x}'\mathbf{x})^{-1}(\mathbf{x}'\mathbf{Y} - c\mathbf{1}) \tag{2.3.1}$$

where $c = \mathbf{1}'(\mathbf{x}'\mathbf{x})^{-1}\mathbf{x}'\mathbf{Y} - 1$, is utilized. Williams tests H_0: f Correct against H_1: f^* Correct using the F statistic with $k-1$ and $n-k+1$ degrees of freedom

$$F = (n-k+1)[\text{RSS}(\bar{f}) - \text{RSS}(\hat{f}^*)/[(k-1)\,\text{RSS}(\hat{f}^*)] \tag{2.3.2}$$

The derivation of F is outlined in Problem 2.5. If H_0 is accepted at level α, that is, $F < F_{1-\alpha,k-1,n-k+1)}$, then the models are considered too close for discrimination. When H_0 is rejected, the model with the least RSS value is selected as correct, but this choice is not quantified (such as by $P(\text{IS}(f_i))$. A test comparing individual models is outlined in Problem 2.6.

The second test based on A2, called the tilde test, was proposed by Atkinson (1969). The null hypothesis tested is that the model's mean square deviations at each location are the same; that is, we test H_0: $\sum_{j=1}^{n}[E(Y_j) - f_j(i)]^2$ Is Constant, for $i = 1, \ldots, k$. To construct this test we first note that H_0 is equivalent to H_0: $\sum_{j=1}^{n} E(Y_j)f_j(i) - \sum_{j=1}^{n} f_j(i)^2/k$ Is Constant, for $i = 1, \ldots, k$. Let $Z(i) = \sum_{j=1}^{n} Y_j f_j(i) - \sum_{j=1}^{n} f_j(i)^2$ for $1 \le i \le k$ and note that under H_0 we have $Z(i) \sim n(c, \sigma^2 \sum_{j=1}^{n} f_j(i)^2)$. With the previous notation ($\mathbf{x}$ consisting of rows formed by the model values), we define $Q = [\mathbf{Z}'(\mathbf{x}'\mathbf{x})^{-1}\mathbf{Z} - (\mathbf{Z}'(\mathbf{x}'\mathbf{x})^{-1}\mathbf{1})^2/(\mathbf{1}'(\mathbf{x}'\mathbf{x})^{-1}\mathbf{1})]$ where $\mathbf{Z}' = (\mathbf{Z}(1),\ldots,Z(n))$ and $\mathbf{1}' = (1,\ldots,1)$. We test H_0 using the F statistic with degrees of freedom $k-1$ and $n-k$:

$$F = (n-k)Q/[(k-1)\,\mathrm{RSS}(\hat{f})] \tag{2.3.3}$$

If F is large, greater than $F_{(1-\alpha,k-1,n-k)}$, we reject H_0. An example of this test is presented in Problem 2.8.

The previous two DMs compare the models of $\mathcal{S}$ with a constructed larger model which has no theoretical modeling basis. Here, the direct contrasting of models is not done. Testing from the point of view of A3, using DMs from C_1 (LQM where $\mathbf{S}(i) = \mathbf{S}$ and $c(i) = 0$ for $1 \le i \le k$), is now considered. With any such LQM, (1.6.1) implies f_i from $\mathcal{S}$ is selected if the event

$$\bigcap_{q \neq i} \{2\mathbf{f}(q,i)'\mathbf{S}\mathbf{R}(i) < \mathbf{f}(q,i)'\mathbf{S}\mathbf{f}(q,i)\} \tag{2.3.4}$$

occurs. Based on the pairwise optimality of LSE, we let $\mathbf{S} = \mathbf{I}$ and (2.3.4) becomes

$$\bigcap_{q \neq l} \{2\mathbf{f}(q,i)'\mathbf{R}(i) < \delta(q,i)\} \tag{2.3.5}$$

where $\delta(q,i) = [\mathbf{f}(q,i)'\mathbf{f}(q,i)]^{1/2}$. The first set of selection probabilities is obtained using Theorem 1.5.1. In connection with LSE, we have from (2.3.5) the conditional bounds

$$P[\mathrm{IS}(f_i)] = \max_{q \neq i} P[Z > \delta(i,q)/(2\sigma)] \tag{2.3.6}$$

and

$$P[\mathrm{CS}(f_i)] = \min_{q \neq i} P[Z > \delta(i,q)/2\sigma)] \qquad (2.3.7)$$

The upper bound on $P(\mathrm{IS})$ given by (1.5.8) can be constructed using (2.2.3). More efficient bounds for these selection probabilities are now sought.

For a fixed $f_i \in \mathcal{S}$, we define a set consisting of two other models of $\mathcal{S}$ by $S_N(i) = \{f_r, f_q : \delta_i(r,q) < 0\}$, where $\delta_i(r,q) = \mathbf{f}(r,i)'\mathbf{f}(q,i)$. Here $S_N(i)$ is not necessarily unique and is chosen according to later considerations.

Remark 2.3.1

In the terminology of Lehman (1966), the models in $S_N(i)$ possess a negative dependence structure in terms of $\mathbf{f}(r,i)'\mathbf{R}(i)$ and $\mathbf{f}(q,i)'\mathbf{R}(i)$. For a further discussion we refer to Eaton (1982).

Lemma 2.3.1

Let $\mathbf{Y}' = (Y_1, Y_2)$ be bivariate normal with zero mean vector and dispersion matrix consisting of ones on the main diagonal and ρ's off the main diagonal. For $c_1 > 0$ and $c_2 > 0$ if $\rho < 0$,

$$P(Y_1 < c_1,\ Y_2 < c_2) \leq P(Z < c_1)P[Z < (c_2 - \rho c_1)/(1-\rho^2)^{1/2}] \qquad (2.3.8)$$

And if $\rho > 0$,

$$P(Y_1 < c_1,\ Y_2 > c_2) \leq P(Z < c_1)P[Z < (c_2 - \rho c_1)/(1-\rho^2)^{1/2}] \qquad (2.3.9)$$

Proof To obtain (2.3.8), we first condition on Y_1, noting that the distribution of Y_2 given $Y_1 = y_1$ is normal with mean ρy_1 and variance $1 - \rho^2$. Thus,

$$P(Y_1 < c_1,\ Y_2 < c_2) = \int_{-\infty}^{c_1} P[Z < (c_2 - \rho y_1)/(1 - \rho^2)^{1/2}]\, dF(Y_1)$$

If $\rho \leq 0$, then on $\{Y_1 < c_1\}$ we have $c_2 - \rho Y_1$ less than $c_2 - \rho c_1$, which implies (2.3.8). Similarly, (2.3.9) is derived by conditioning on Y_1.

Theorem 2.3.1

Let $k \geq 3$ and assume normality of the error. Let f_i be chosen using LSE where $\delta(i,r) < \delta(i,q)$ for $q \neq r$. Further, assume there exists at least one model f_q, from $\mathcal{S}$, such that $S_N(i) = \{f_r, f_q\}$. Then

$$P[\mathrm{IS}(f_i)] \leq \min P[Z > \delta(r,i)/(2\sigma)] \times P\left\{Z \leq [\Delta(r,q) + \rho\delta(q,i)]/[2\sigma(1 - \rho^2)^{1/2}]\right\} \quad (2.3.10)$$

and

$$P[\mathrm{CS}(f_i)] \leq \min P[Z \leq \delta(r,i)/(2\sigma)] \times P\left\{Z \leq [\delta(q,i) - \rho\delta(r,i)]/[2\sigma(1 - \rho^2)^{1/2}]\right\} \quad (2.3.11)$$

where $\Delta(r,q) = [\delta(q,i)^2 - 2\delta_i(q,r)]/\delta(q,i)$, $\rho = \delta_i(r,q)/[\delta(r,i)\delta(q,i)]$, and the minimums are taken over all f_q comprising $S_N(i)$.

Proof To derive (2.3.11) we use (2.3.8) with $Y_1 = \mathbf{f}(r,i)'\mathbf{R}(i)/[\delta(r,i)\sigma]$, $c_1 = \delta(r,i)/(2\sigma)$, $Y_2 = \mathbf{f}(q,i)'\mathbf{R}(i)/[\delta(q,i)\sigma]$, and $c_2 = \delta(q,i)/(2\sigma)$. To show (2.3.10) we note that, from (2.3.5), f_i is selected over f_r if $2\mathbf{f}(i,r)'\mathbf{R}(r) > \delta(r,i)^2$ and over f_q if $2\mathbf{f}(q,i)'\mathbf{R}(i) < \delta(q,i)^2 - 2\delta_i(q,r)$ (letting $\mathbf{R}(i) = \mathbf{R}(r) + \mathbf{f}(r,i)$). Assuming f_r is correct, we apply (2.3.9) where Y_1, c_1, and Y_2 are defined as above and c_2 is $\Delta(r,q)/(2\sigma)$.

When there exists models to form $S_N(i)$ the conditional IS probability bound (2.3.10) is an improvement over (2.3.6). With this bound we can test A3 as introduced in Section 1.5. The upper bound, although in the wrong direction, on $P[\text{CS}(f_i)]$ can be utilized when k is small.

Example 2.3.1

Let S consist of four models defined by model functions $f(1) = 1.25 - X_2, f(2) = 1 - x, f(3) = \exp(-x)$, and $f(4) = 1 - x/2$. One observation at each of three locations, x values, is taken. The model values are given in Table 2.3. On graphing the models we find that the "outermost" models are f_1 and f_2. Under the possible model selections, conditional bounds (2.3.11) and (2.3.10) are calculated using $\sigma = .2$. These are listed in Table 2.4. For example, if f_4 is chosen using LSE we have $S_N(4) = \{f_1, f_3\}$ where $\delta(3,4) = .195 < \delta(q,4)$ for $q = 1, 2$. Also $\delta(1,4) = .433$ and $\delta(1,3) = -.00286$ imply $\rho = -.00286/(.03806 \times .1875)$. From (2.3.10), $P[\text{IS}(f_4)] \leq P(Z > .4877)P(Z < 1.5064) = .2926$. Similarly, from (2.3.11), $P[\text{CS}(f_4)] \leq P(Z \leq .4877)P[Z \leq 1.0997) = .5939$. We note that models f_1 and f_2 have $S_N = \varnothing$. Here, over the models in S, $\max P[\text{IS}(f_i)] \leq .3117$. Based on question A3, the modeler would accept H_0: Not Enough Information to Make a Selection due to the large conditional IS probabilities.

Table 2.3 Four-Model Example—Location Values

Responses	f_1	f_2	f_3	f_4
0.0	1.25	1.00	1.0000	1.00
0.5	1.00	.50	.6065	.75
1.0	.25	.00	.3678	.50

Table 2.4 Probability Bounds

	Selected Model			
	f_1	f_2	f_3	f_4
$P[\text{IS}(f_i)]$	.1305	.1692	.3117	.2926
$P[\text{CS}(f_i)]$	.8605	.8308	.6844	.5939

Using LSE, the inside models, or models between other models at most locations, have a difficult time being chosen even if they are correct. This is seen in Example 2.3.1 in the lowering of the upper bound on $P[\text{CS}(f_i)]$ for the inside models f_3 and f_4 relative to models f_1 and f_2. If inside models are to compete with their rivals, then DMs other than simultaneous ones may be needed.

In practice, when σ^2 is unknown, an estimate such as (2.2.6) can be employed. Noting Example 2.2.2, the bias encountered in substituting $\hat{\sigma}$ for σ should be relatively small. For this reason, throughout the remaining text we assume that the modeler may freely substitute $\hat{\sigma}$ for σ in DFs.

2.4 SEQUENTIAL METHODS

Under the viewpoint of A3, DMs which consist of a series of model comparisons are constructed. The comparisons are referred to as steps in the procedure. At each step a model under consideration is fixed and a QDF, defined by (1.4.3), is utilized to either select as the true one or eliminate from competition the model in question. The DM continues until a model is chosen. These sequential methods, denoted SM, were first proposed by Borowiak (1983) and allow for easier selection, in comparison to simultaneous, of "inside" models.

On step i, suppose we have $s = k - i + l$ possible completely determined models contained in set $\mathcal{S}_i$. Thus, $\mathcal{S}_i = \{s$ proposed models from $\mathcal{S}\}$ and $\mathcal{S}_i \subseteq \mathcal{S}$. We select $f_{(i)}$ from $\mathcal{S}_i$ and consider LQMs which take $\mathbf{S}_i = \mathbf{A}_i\mathbf{A}_i'$ where $\mathbf{A}_i$ is $n \times 1$. From (2.3.4) we note that $f_{(i)}$ is selected over all rivals in $\mathcal{S}_i$ if the event

$$\bigcap_{r \in B_{i,1}} \{2\mathbf{A}_i'\mathbf{R}(i) < \mathbf{f}(r,(i))'\mathbf{A}_i(\mathbf{A}_i'\mathbf{A}_i)^{1/2}\}$$

$$\bigcap \bigcap_{r \in B_{i,2}} \{2\mathbf{A}_i'\mathbf{R}(i) \geq \mathbf{f}(r,(i))'\mathbf{A}_i/(\mathbf{A}_i)^{1/2}\} \tag{2.4.1}$$

occurs, where $B_{i,1} = \{r : \mathbf{f}(r,(i))'/\mathbf{A}_i \geq 0\}$ and $B_{i,2} = \{r : \mathbf{f}(r,(i))'\mathbf{A}_i < 0\}$. For a fixed $f_{(i)}$ and f_r from $\mathcal{S}_i$, $r \neq (i)$, let

$$d_{(i)}(r) = \mathbf{f}(r,(i))'\mathbf{A}_i/(\mathbf{A}_i'\mathbf{A}_i)^{1/2} \tag{2.4.2}$$

and define

$$b_i = \begin{cases} \min_{Y \in B_{i,1}} \{d_{(i)}(r)\} & \text{if } B_{i,1} \neq \varnothing \\ \text{and } \infty & \text{if } B_{i,1} = \varnothing \end{cases} \tag{2.4.3}$$

and

$$a_i = \begin{cases} \max_{Y \in B_{i,2}} \{d_{(i)}(r)\} & \text{if } B_{i,2} \neq \varnothing \\ \text{and } -\infty & \text{if } B_{i,2} = \varnothing \end{cases} \tag{2.4.4}$$

The general SM is now presented in the form of a theorem.

Theorem 2.4.1

Let the models of $\mathcal{S}$ be completely determined where normality of the error is assumed. Define an SM as follows; on step i choose $f_{(i)} \subseteq \mathcal{S}_i$ and stop and declare $f_{(i)}$ correct if it is selected based on the LQMs using $\mathbf{S}(r) = \mathbf{A}_i\mathbf{A}_i'$ for $r \neq (i)$ $(k - i + 1$ comparisons); that is, we select $f_{(i)}$ if

$$a_i \leq 2\mathbf{A}_i'\mathbf{R}(i)/(\mathbf{A}_i'\mathbf{A}_i)^{1/2} \leq b_i \tag{2.4.5}$$

where a_i and b_i are defined by (2.4.3) and (2.4.4), respectively. Otherwise, drop $f_{(i)}$ out of consideration, forming $\mathcal{S}_{i+1}$, and proceed to step $i + 1$. If step $k - 1$ is reached, the selection is made between the remaining two models, denoted $f_{(k-1)}$ and $f_{(k)}$. Let $\gamma_i = \max\{|d_{(i)}(r)|/(2\sigma), f_r \in \mathcal{S}_i, r \neq (i)\}$, $\alpha_i = \min\{|a_i|/(2\sigma), b_i/(2\sigma)\}$, and $\beta_i = \max\{|a_i|/(2\sigma), b_i/(2\sigma)\}$ for $1 \leq i \leq k_1$. Also, suppose

$$\mathbf{A}_i'\mathbf{A}_r = 0 \qquad \text{for } i \neq r \tag{2.4.6}$$

If $f_{(1)}$ is selected

$$P[\mathrm{CS}(f_{(1)})] = P(-\alpha_1 \leq Z \leq \beta_1) \tag{2.4.7}$$

and for $i \geq 2$

$$P[\mathrm{IS}(f_{(1)} \,|\, f_{(i)})] \leq P(\alpha_1 \leq Z \leq \beta_1 + 2\alpha_1) \tag{2.4.8}$$

On the other hand, if $f_{(1)}$ is correct

$$P[\mathrm{IS}(f_{(i)} \,|\, f_{(1)})] \leq P(Z > \alpha_1) + P(Z > \beta_1) \tag{2.4.9}$$

for $i \geq 2$. Further, for $2 \leq i < j$

$$P[\text{IS}(f_{(i)} \mid f_{(j)})] \leq P(\alpha_i < Z < \beta_i + 2\alpha_i) \times \prod_{r=1}^{i-1} [P(Z \leq 2\gamma_r - \alpha_r) = P(Z > \beta_r + 2\gamma_r)] \qquad (2.4.10)$$

and for $i > j \geq 2$

$$P[\text{IS}(f_{(i)} \mid f_{(j)})] \leq [P(Z > \alpha_j) + P(Z > \beta_j)] \times \prod_{r=1}^{j-1} [P(Z \leq 2\gamma_r - \alpha_r) + P(Z > \beta_r + 2\gamma_r)] \qquad (2.4.11)$$

In addition, for $2 \leq i \leq k-1$

$$P[\text{CS}(f_{(i)})] \geq P(-\alpha_i \leq Z \leq \beta_i) \times \prod_{j=1}^{i-1} [P(Z < \alpha_j) + P(Z > \beta_j + 2\alpha_j)] \qquad (2.4.12)$$

and $P[\text{CS}(f_{(k)})] = P[\text{CS}(f_{(k-l)})]$.

Proof From (2.4.1) we see that step decisions are based on events (2.4.5). Condition (2.4.6) implies independence among step decisions. Under various combinations of hypothesized and selected models, (2.4.7)–(2.4.12) are derived by standardization similar to that done in Theorem 2.2.1.

Corollary 2.4.1

Under the assumptions and SM of Theorem 2.4.1, let $p_1 = P(Z > \alpha_1) + P(Z > \beta_1)$, $t_1 = P(\alpha_1 \leq Z \leq \beta_1 + 2\alpha_1)$, and for $2 \leq i \leq k-1$ define

$$p_i = [P(Z > \alpha_I) + P(Z > \beta_i)]$$

$$\times \prod_{j=1}^{i=1} [P(Z \leq 2\gamma_j - \alpha_j) + P(Z > \beta_j + 2\gamma_j)]$$

and

$$t_i = P(\alpha_i \leq Z \leq \beta_i + 2\alpha_i)$$
$$\times \prod_{j=1}^{j-i} [P(Z \leq 2\gamma_j - \alpha_j) + P(Z > \beta_j + 2\gamma_j)]$$

Here,

$$P[\mathrm{IS}(f_{(1)})] \leq t_1 \tag{2.4.13}$$

and for $2 \leq i \leq k-1$

$$P[\mathrm{IS}(f_{(i)})] \leq \max\{p_1, \ldots, p_{i-1}, t_i\} \tag{2.4.14}$$

Proof The proof is a direct consequence of (2.4.7)–(2.4.12).

The general SM allows for many varieties through the choices of $\mathbf{A}_i$ according to stipulation (2.4.6). In the balance of this section two SMs are studied. The first yields sharp calculations of selection probabilities, whereas the second searches for near-optimal step constructions.

Following some notation, the first SM is derived through a series of lemmas. On step i fix $f_{(i)} \in \mathcal{S}_i$ where $\mathcal{S}_i = \{f_1, \ldots, f_s\}$ and define the $n \times (s-1)$ matrix (here $s = k - i + 1$).

$$\mathbf{G}_i' = (\mathbf{f}(1,(i)), \ldots, \mathbf{f}((i)-1,(i)), \mathbf{f}((i)+1,(i)) \ldots, \mathbf{f}(s,(i))) \tag{2.4.15}$$

In (2.4.15) we assume $s \leq n+1$ and $\mathbf{G}_i$ has rank $s-1$.

Lemma 2.4.1

For $1 \leq i \leq k-1$ define

$$\mathbf{A}_i = \mathbf{G}_i'(\mathbf{G}_i\mathbf{G}_i')^{-1}\mathbf{1} \tag{2.4.16}$$

where $\mathbf{1}' = (1,\ldots,1)$. For any $f_q \in \mathcal{S}_i$, $q \neq (i)$,

$$\mathbf{f}(q,(i))'\mathbf{A}_i = 1 \tag{2.4.17}$$

and (from (2.4.2))

$$d_{(i)}(q) = [\mathbf{1}'(\mathbf{G}_i\mathbf{G}_i')^{-1}\mathbf{1}]^{-1/2} \tag{2.4.18}$$

Further, (2.4.18) is a maximum over the set of values of (2.4.2) restricted to choices of $\mathbf{A}_i$ in which (2.4.17) holds.

Proof We first represent the constraints in (2.4.17) by the system $\mathbf{G}_i\mathbf{A}_i = \mathbf{1}$. Since $\mathbf{G}_i$ has full row rank, solutions of the form $\mathbf{A}_i = (\mathbf{G}_i'\mathbf{G}_i)^{-}\mathbf{G}_i'\mathbf{1}$ exist where $(\mathbf{G}_i'\mathbf{G}_i)^{-}$ is a generalized inverse of $(\mathbf{G}_i'\mathbf{G}_i)$. The quantity $\mathbf{A}_i'\mathbf{A}_i$ is minimized by any least squares inverse, as defined by Winer (1979, pp. 77–79). The Moore–Penrose inverse is a least squares inverse and applying the form of Lowere (1982) is $\mathbf{G}_i'(\mathbf{G}_i\mathbf{G}_i')^{-1}(\mathbf{G}_i\mathbf{G}_i')^{-1}\mathbf{G}_i$. Substitution of this form completes the proof.

The common value of $d_{(i)}(q)$, $q \neq (i)$, given by (2.4.18), being dependent only on the choice of $f_{(i)}$, is denoted d_i for $i = 1, \ldots, k-1$.

Lemma 2.4.2

In an SM for steps i and j, $i > j$, we note $\mathcal{S}_i \subseteq \mathcal{S}_j$ and define $\mathbf{A}_i$ and $\mathbf{A}_j$ by (2.4.16). Then $\mathbf{A}_i'\mathbf{A}_j = 0$.

Proof From Lemma 2.4.1, d_j being a maximum implies $\mathbf{A}_i$ takes the form

$$\mathbf{A}_i = \sum \lambda_r f(r,(i)) \tag{2.4.19}$$

where the λ_r's are constants and the summation is over $r \geq 1, f_r \in S_j$, and $r \neq (i)$. For $f(r,(i))$ in (2.4.19) we substitute $f(r,(j)) - f((i),(j))$. The proof is completed using (2.4.17).

Lemma 2.4.3

Denote the SM outlined in Theorem 2.4.1 obtained by defining $\mathbf{A}_i$, $1 \leq i \leq k-1$, by (2.4.16) as SM1. With SM1, (2.4.6) holds, $\beta_i = \infty$, and $\gamma_i = \alpha_i$ for $i = 1, \ldots, k-1$.

Proof From Lemma 2.4.2 we see that (2.4.6) holds. The proof is completed noting that (2.4.17) implies $B_{i,2} = \varnothing$ and Lemma 2.4.1 yields $\alpha_i = \gamma_i$ for $1 \leq i \leq k-1$.

Theorem 2.4.2

We apply SM1 under the conditions of Theorem 2.4.1 and define

$$p_1 = P[Z > d_1/(2\sigma)] \tag{2.4.20}$$

For $2 \leq s \leq k-1$

$$p_s = P[Z > d_s/(2\sigma)] \prod_{J=1}^{s-1} P[Z \leq d_j/(2\sigma)] \tag{2.4.21}$$

and $p_k = p_{k-1}$. Then

$$P[\mathrm{IS}(f_{(i)}) = \max\{p_1, \ldots, p_i\} \tag{2.4.22}$$

and

$$P[\mathrm{CS}(f_{(i)})] = \prod_{j=1}^{i} P[Z \le d_j/(2\sigma)] \tag{2.4.23}$$

Also, $P(\mathrm{CS})$ defined by (1.5.5) is the minimum over $f_{(i)}$, $1 \le i \le k-1$, of $P[\mathrm{CS}(f_{(i)})]$.

Proof The proof is a consequence of Lemma 2.4.2 and noting $\alpha_i = d_i/(2\sigma)$ for $i = 1, \ldots, k-1$.

Example 2.4.1

The SM1 procedure is applied to the set of models introduced in Example 2.3.1 (model function values listed in Table 2.3). The choices $f_{(i)} = f_i$ for $i = 1, \ldots, 4$ are employed. On the first step, from (2.4.15)

$$\mathbf{G}' = \begin{pmatrix} -.25 & -.25 & -.25 \\ -.5 & -.393 & -.25 \\ -.25 & .1178 & .25 \end{pmatrix}$$

which leads to

$$(\mathbf{G}\mathbf{G}')^{-1} = \begin{pmatrix} 76.170 & -206.151 & 158.477 \\ & 584.233 & -455.602 \\ & & 362.148 \end{pmatrix}$$

From (2.4.16)

$$\mathbf{A}_1 = \mathbf{G}_1' \begin{pmatrix} 28.4961 \\ -77.5194 \\ 65.4961 \end{pmatrix} = \begin{pmatrix} -4 \\ 0 \\ 0 \end{pmatrix}$$

Hence, $d_1 = .25$. Listed in Table 2.5 are $\mathbf{A}_i$ and d_i for $1 \leq i \leq 3$. Using Theorem 2.4.2 with $\sigma = .2$ we find $p_1 = .266$, $p_2 = .103$, and $p_3 = .155$. Thus we observe $P[\text{IS}(f_i)] = .266$ for $a \leq i \leq k$, $P[\text{CS}(f_2)] = .632$, and $P[\text{CS}(f_4)] = .476$. Comparing these results to the bounds for LSE listed in Table 2.4, we note that SM1 is superior when f_4 is selected (lower conditional IS probability bound) but inferior to LSE (higher $P[\text{IS}(f_1)]$) if f_1 is chosen. This supports the notion that SM1 is more useful in discriminating among inside models. Also, comparing $P[\text{CS}(f_i)]$ for SM1 to the upper bound (2.3.11) associated with LSE, we find that the sequential method is not desirable when selecting an outside model.

In the second SM, which we denote SM2, we attempt to (nearly) optimize selection probabilities at each step under condition (2.4.6). In this way a more powerful SM, over SM1, is constructed when the true model occurs in early steps, that is, larger $P[\text{CS}(f_{(i)})]$ for small i. This gain can also be realized over LSE. For example, if $f_{(1)}$ is the true model the event of a correct selection using LSE is the intersection of $k - 1$ events given by (2.3.5). On the other hand, the correct choice utilizing an SM corresponds to the occurrence of a single event, namely (2.4.5) with $i = 1$. This comparison is particularly cogent in the construction of tests based on question A1 of the form (1.5.9).

Optimality in terms of selection probabilities for SMs is not easily attained; therefore we search for a near-optimal construction.

Table 2.5 SM1 Procedure—$f_{(i)} = f_i$

	d_i	$\mathbf{A}_i'$		
Step 1	.2500	−4.000	0.000	0.000
Step 2	.4328	0.000	−3.416	3.708
Step 3	.1951	0.000	3.769	3.473

On step i the choice of $\mathbf{A}_i$ of Theorem 2.4.1 is based on maximizing a function of α_i for $1 \leq i \leq k-1$. For a fixed $f_{(i)} \in \mathcal{S}$ we maximize

$$\Delta_i = \prod_{j \neq i} d_{(i)}(j) \tag{2.4.24}$$

where (2.4.6) holds and $d_{(i)}(j)$ is given by (2.4.2). On step i denote a vector of weights by $\mathbf{W}_i' = (w_{i,1}, \ldots, w_{i,s})$ where $s = k - 1 + 1$. Further, let $\mathbf{W}_i^{*\prime} = (1/w_{i,1}, \ldots, 1/w_{i,s})$ for $i = 1, \ldots, k-1$. The near-optimal SM is now presented.

Theorem 2.4.3

Under the conditions of Theorem 2.4.1, the SM which maximizes (2.4.24) on each step subject to constraint (2.4.6), denoted SM2, takes $\mathbf{A}_i$ values defined according to the following conditions: on step $i = 1$ let

$$\mathbf{A}_1 = \mathbf{G}_1'\mathbf{W}_1 \tag{2.4.25}$$

where $\mathbf{W}_1$ is a solution to $\mathbf{W}_1^* = \mathbf{G}_1\mathbf{G}_1'\mathbf{W}_1$, and for step i, $i \geq 2$, define

$$\mathbf{A}_i = \mathbf{G}_i\mathbf{W}_i - \sum_{j=1}^{i-1} q_j \mathbf{A}_j \tag{2.4.26}$$

where $q_j = \mathbf{A}_j'\mathbf{G}_j\mathbf{W}_i/\mathbf{A}_j'\mathbf{A}_j$ and $\mathbf{W}_i$ is a solution to $\mathbf{W}_i^* = \mathbf{G}_i\mathbf{G}_i'\mathbf{W}_i - \sum q_j \mathbf{G}_i \mathbf{A}_j$.

Proof First, from (2.4.24) we have

$$\ln(\Delta_1) = 2 \sum_{j \neq (i)} \ln[\mathbf{A}_1'\mathbf{f}(j, (i))] - (k-1)\ln(\mathbf{A}_1'\mathbf{A}_1) \tag{2.4.27}$$

Setting the derivative of $\ln(\Delta_1)$ equal to zero yields the solution $\mathbf{A}_1 = \sum c_j \mathbf{f}(j,(i))$ where $c_j = c/(\mathbf{A}_1'\mathbf{f}(j,(i))$ and c is a constant. Since the value of c is immaterial, setting $c = 1$ implies $\mathbf{A}_1 = \mathbf{G}_1'\mathbf{W}_1$ where $\mathbf{W}_1' = (1/\mathbf{A}_1'(\mathbf{f}(1,(i)),\ldots,1/\mathbf{A}_1'\mathbf{f}(k,(i)))$. Noting (2.4.15), we see $\mathbf{W}_1^* = \mathbf{G}_1\mathbf{A}_1 = \mathbf{G}_1\mathbf{G}_1'\mathbf{W}_1$, which gives (2.4.25). The derivation of (2.4.26) follows a similar approach where $\ln(\Delta_i) + \sum_{j=1}^{i_1} \lambda_j A_i' A_j$ is maximized.

To employ SM2, W_i, $1 \leq i \leq k-1$, which solves the appropriate equation needs to be found. The solution is generally unique (see Problem 2.12) and may be found by the method outlined in the following remark.

Remark 2.4.1

A possible procedure for finding the $\mathbf{W}_i$'s of Theorem 2.4.3 is to solve an iterative system. For initial $\mathbf{W}$ we find, for $i = 1$,

$$\mathbf{V} = \mathbf{G}_1\mathbf{G}_1'\mathbf{W} \tag{2.4.28}$$

and for $i \geq 2$

$$\mathbf{V} = \mathbf{G}_i\mathbf{G}_i'\mathbf{W} - \sum_{j=1}^{i-1} q_j \mathbf{G}_i \mathbf{A}_j \tag{2.4.29}$$

where $q_j = \mathbf{A}_j'\mathbf{G}_i,\mathbf{W}/\mathbf{A}_j'\mathbf{A}_j$ for $1 \leq j \leq i-1$. In either (2.4.28) or (2.4.29) we let $\mathbf{V}' = (v_1,\ldots,v_{s-1})$ and take as the next vector (in place of $\mathbf{W}$) $(c/v_1, \ldots, c/v_{s-1})$ where $c = [1/\sum(1/v_i^2)]^{1/2}$. We continue iteratively until convergence, as measured by the change in successive $\mathbf{W}$'s, is attained.

The strength of SM2 lies in its ability to discriminate in early steps. For this reason this DM is particularly suited to the construction of tests of the form (1.5.9) based on viewpoint A1.

Example 2.4.2

The modeler is willing to assume a particular model, say f_i, is correct unless the observed data forces disbelief. Here SM2 is applied with $f_{(1)} = f_i$ to test

$$H_0: E(Y_j) = f_j(i) \qquad \text{for } 1 \le j \le n$$

against

$$H_1: E(Y_j) = f_j(q) \qquad \text{for } 1 \le j \le n \text{ and some } f_q \in S,\ q \ne i$$

The critical region comes from the decision of step one of SM2. We reject H_0 if either $2\mathbf{A}_1'\mathbf{R}(i)/(\mathbf{A}_1'\mathbf{A}_1)^{1/2} > b_1$ or $2\mathbf{A}_1'\mathbf{R}(i)/(\mathbf{A}_1'\mathbf{A}_1)^{1/2} < a_1$ where $\mathbf{A}_1$ is given by (2.4.25) and a_1 and b_1 are defined by (2.4.4) and (2.4.3), respectively. The observed significance level for this test is p_1 given in Corollary 2.4.1.

Through SM powerful DMs with conditional selection probabilities or appropriate bounds are constructed. Based on Atkinson's questions, namely A1 or A3, tests can be formed with the help of SM (see Example 2.4.3 and Problem 2.11). For initial steps SM2 results in smaller conditional probabilities of IS than SM1. This is due to the restrictive nature of condition (2.4.17) when SM1 is employed. However, on later steps SM1 enjoys a gain in the conditional probabilities over SM2. The question of which DM, either

Table 2.6 SM2 Procedure—$f_{(i)} = f_i$

	a_i	b_i	α_i	β_i	γ_i
Step 1	.3790	$-\infty$	.9475	∞	1.2835
Step 2	.3807	$-\infty$	.9581	∞	1.3548
Step 3	.0481	$-\infty$	1.2025	∞	1.2025

an SM like SM1 or SM2 or a simultaneous procedure such as LSE, is addressed in the next section.

As in the case of LSE, we have focused our attention on conditional selection probabilities. This is because $P(\text{IS})$ is hard to calculate and upper bounds similar to (1.5.8) tend to be too conservative. In the case of SM1

$$P(\text{IS}) \leq \sum_{j=1}^{k-1} p_i \tag{2.4.30}$$

where p_i is defined by (2.4.20) and (2.4.21). While (2.4.30) does assess the SM, it may be too large to be used in hypothesis testing. As mentioned in the next section, this measure is of use in the construction of efficient SM1 methods.

2.5 SELECTION OF DISCRIMINATION METHOD

In most applications the researcher wishes to compare rival completely determined models which are based on theoretical derivations. For this reason, assuming the correct model exists and is in $\mathcal{S}$, question A3 is considered and a choice of DM must be made. Open to the modeler are the simultaneous methods, of which LSE is the most common, along with the SMs, namely SM1 and SM2. This choice is compounded by the fact that in conjunction with SMs designations are needed for $f_{(i)}$, $1 \leq i \leq k-1$. In this section our main objective is to present an efficient criterion for the selection of a DM to be used in the case of completely determined models. Any such assessment is to be based on selection probabilities or appropriate bounds.

We search for a DM from a non-Bayesian viewpoint. By this we mean no prior information on the correctness of any candidate model is utilized. If prior probabilities (of true models) are present, then we need only condition on these probabilities in the previous method's DFs (see Problem 2.13).

From the class of LQMs only LSE is considered, whereas both SMs, namely SM1 and SM2, are taken as possibilities. The initial method we might entertain is to compare conditional IS probabilities for the DMs. For each DM we note the model selected and find the conditional probability, or corresponding bound, of an IS. Here, the SMs employ $f_{(i)}$, $1 \leq i \leq k-1$, which minimize $P[\text{IS}(f_{(i)})]$ where $f_{(i)}$ is selected. This assessment of the DMs is dynamic in nature in that it depends on the empirical data through the model which is selected. For this reason there are two main drawbacks. First, inside models, since they are harder to select (especially for LSE), are at an unfair disadvantage. Second, conditional selection probabilities in this setting are hard to calculate. In fact, the overall estimate of the conditional IS probability given by $\min P[\text{IS}(f_i)]$ can be an overestimate (Problem 2.14). Hence, a more stable DM assessment procedure is desired.

The value $P(\text{IS})$, as defined by (1.5.4), is independent of the model selected. However, the upper bounds on this quantity for LSE, namely (1.5.8) using (2.2.3) when $k \geq 3$, and for SM1, given by (2.4.30), can be very conservative. For this reason this measure would not, in general, be a good evaluator of prospective DMs.

Under any hypothesized true model the power of any DM is entailed in its CS probabilities. For a proposed DM the values $P[\text{CS}(f_i)]$, $1 \leq i \leq k$, measure the strength of the method in the same way as the power function of a hypothesis test indicates its ability to discriminate between hypotheses. This philosophy indicates a suitable criterion to be used to assess the

DMs. Since the true model is unknown, we select the DM which maximizes $P(\mathrm{CS})$ as defined by (1.5.5). For any SM we choose $f_{(i)}$, $1 \leq i \leq k-1$, based on the same criterion; that is, select $f_{(i)}$ which maximizes $P(\mathrm{CS})$. In the LSE case only upper bounds on $P[\mathrm{CS}(f_i)]$ exist and, therefore, further analysis in the form of a simulation evaluation may be required (see Example 2.5.1). After choosing a DM based on Atkinson's question A1 or A3, tests can be constructed as in Examples 2.2.1 and 2.4.3.

Example 2.5.1

The data and proposed models introduced in Example 2.3.1 are investigated. The proposed DM criterion, maximization of $P(\mathrm{CS})$, is used to select a DM. The SM1 procedure which maximizes the minimum probability of a correct selection takes $f_{(1)} = f_1$, $f_{(2)} = f_4$, $f_{(3)} = f_2$, and $f_{(4)} = f_3$. From (2.4.18) we calculate $d_1 = .25$, $d_2 = .4808$, and $d_3 = .3829$. The other SM, namely SM2, utilizes choices $f_{(1)} = f_2$, $f_{(2)} = f_1$, $f_{(3)} = f_4$, where $\beta_i = 0$ for $1 \leq i \leq 3$, $\alpha_1 = .8935$, $\alpha_2 = 1.0298$, and $\alpha_3 = .0120$. For these SMs lower bounds on $P[\mathrm{CS}(f_i)]$, $1 \leq i \leq 4$, are found utilizing (2.4.23) and (2.4.7) or (2.4.12). To compare these SMs with LSE, noting that only upper bounds on CS exist for LSE in the form of (2.3.7) and (2.3.11), simulation computations are utilized. Using the model selection regions given by (2.3.5), the observed proportion of CS out of 5000 repetitions, taking normal and independent errors with $\sigma = .2$, is found. The results for these three DMs are listed in Table 2.7.

The SM1 procedure is chosen with least favorable case where f_3 is the true model. Here $P(\mathrm{CS}) = .5399$. The LSE procedure is comparable to the SM1 method in terms of the maximization of $P(\mathrm{CS})$. However, since simulation evaluation is required for LSE, the SM is preferred.

Table 2.7 Comparison of $P(\mathrm{CS}(f))$—True Model

	f_1	f_2	f_3	f_4
SM1	.7340	.5399	.5399	.6498
SM2	.7001	.8142	.3834	.3834
LSE	.8230	.8280	.5320	.5680

We remark, as pointed out by this example, that SM1 has an advantage over SM2 and LSE in that it is superior in discriminating among inside models. If only two or three models are of interest, then the SM2 method, which is based on near optimality on early steps, may be preferred. This is the case when testing a preassigned model.

When S contains a large number of models other methods for determining an efficient DM may be required. For even moderate k the number of possible DMs can grow large. For instance, with any SM the number of possible models resulting from various choices of $f_{(i)}$, $1 \le i \le k-1$, is $k!/2$. Here, the pruning of the number of models in S using a fit statistic, such as (2.1.1), is helpful. A large number of possible DMs may prohibit the utilization of the maximum $P(\mathrm{CS})$ criterion. From the previous discussion, when this number is large a reasonable approach, under question A3, is to apply SM1. On each step the model function values are graphed and the "outermost" model is selected for $f_{(i)}$. Also, in some cases upper bound (2.4.30) can be used to trim the number of SM1 methods. Here, if a large value of p_i occurs, for any $i \ge 1$, the SM1 procedure, in terms of choices for $f_{(i)}$, may be discarded.

2.6 AN APPLICATION

The purpose of this section is to demonstrate in a real experimental setting the previously proposed discrimination techniques. Since we are selecting among competing completely determined models, no unknown parameters in the models exist. Thus, if unknown parameters are present they must be estimated from a source independent of the observed data. In addition, estimators used should be the result of a large enough set of data to ensure convergence (see Chapter 4).

The settling of heavy metal, in particular mercury, into sediment due to pollution in a lake is explored. Four models describing the concentration of a metal with depth in sediment are constructed by Borowiak and Walters (1978). These models account for biological mixing, varying metal input concentrations with time, and diffusion. The models include only a sedimentation rate parameter to be estimated. They take on the following characteristics: f_1, constant input and no diffusion; f_2, constant input with diffusion; f_3, variable input and no diffusion; and f_4, variable input with diffusion. These models are applied to sediment samples taken from the western basin of Lake Erie, where the sedimentation rate is estimated from an outside source that consists of many sample sediments. The amount of mercury (p.p.m.) with depth predicted by these models, along with the observed data, is given in Table 2.8. We remark that the repeat observations at each depth result from different samples being drawn from the same vicinity.

The scientist, a geologist, assumes the true model is in S and takes up question A3, namely asking, is there enough evidence to make a selection? But, as mentioned in the previous section, before considering a DM we try to eliminate highly unlikely models.

Table 2.8 Sedimentation Models

Depth	Responses	f_1	f_2	f_3	f_4
2	2.37, 2.10	2.31	2.19	2.09	1.94
7	2.01, 1.95	1.89	1.91	1.79	1.77
11	1.00, 1.07	1.26	1.35	1.35	1.40
15	.63, .87	.70	.80	.97	1.04
22	.33, .19	.16	.21	.51	.58
26.5	.11, .07	.03	.07	.34	.39

To assess the model's fit to the responses we utilize the variance estimate based on repeat samples given by (2.2.6). Here we have two repetitions at each of six locations ($r_j = 2$ for $j = 1, \ldots, 6$). A statistic, which is an extension of (2.1.1) for estimated σ^2, to test for lack of fit of model f_i, under correctness, is

$$\text{LOF}(f_i) = P[F_{(n,m)} > \text{rss}(f_i)/(n\hat{\sigma}^2)] \qquad (2.6.1)$$

where, as before, there are n data points, $\text{rss}(f_i)$ is the observed RSS due to model f_i, and $m = \sum_{j=1}^{s}(r_j - 1)$. In (2.6.1) we note that $F_{(n,m)}$ is an F random variable with n and m degrees of freedom. In particular, we find that $\hat{\sigma}^2$ is .0267 and for each model (2.6.1) is calculated. These values along with the RSS are listed in Table 2.9. Only model f_4 is found to have a serious lack of fit and is eliminated from consideration. Hence, the set of competing models is $\mathcal{S} = \{f_1, f_2, f_3\}$.

The SM1 procedure suggested in Section 1.5 is utilized for discrimination. Among choices of $f_{(i)}$, $1 \le i \le k-1$, the one which maximizes $P(\text{CS})$ is found. Here $f_{(1)} = f_3$, $f_{(2)} = f_1$, and $f_{(3)} = f_2$ and the step-by-step calculations are listed in Table 2.10. Thus, on the first step $2\mathbf{A}'\mathbf{r}(1) = 1.137 > 1$ eliminates f_3, and on step two

Table 2.9 Lack of Fit

	f_1	f_2	f_3	f_4
RSS	.241	.303	.740	1.162
F	.752	.946	2.310	3.627
LOF	.683	.562	.157	.062

$2\mathbf{A}'\mathbf{r}(2) = -.061 < 1$ results in the selection of f_1. The conditional model selection probability bounds are obtained by substituting $\hat{\sigma} = (.0267)^{1/2}$ in (2.4.20)–(2.4.23) and assuming a small amount of bias (see Section 2.2). We find $P[\text{IS}(f_1)] = .195$. Further, the various CS probabilities (leading to this SM) are $P[\text{CS}(f_3)] = .964$, $P[\text{CS}(f_1)] = .751$, and $P[\text{CS}(f_2)] = .751$, which imply $P(\text{CS}) \geq .751$.

For this data the model constructed with constant mercury input and no diffusion is selected with some degree of confidence; that is, the conditional probability of an IS while selecting f_1 is estimated as less than .2. This result is not too bad if we consider the equally likely model case where this probability is 1/3. At this marginal level we assume there is enough evidence to make a selection. The geologist is now left to interpret this finding. Since the main source of pollution was halted some time ago, but f_1 is chosen over f_3, the effects of the pollution appear to be long-lasting due to biological mixing.

Table 2.10 SM1 Procedure

	$f_{(i)}$	$2\mathbf{A}_i'\mathbf{e}_i$	d_i	Decision
Step one	f_3	1.137	.589	Eliminate f_3
Step two	f_1	−.061	.272	Select f_1

Remark 2.6.1

In the original source for this study there existed no repeat data points for any core sample. We stress the fact that a lack of repeat data points hinders a complete analysis since the variance in the responses cannot be estimated outside the proposed models.

The elimination of unlikely models reduces S to a more manageable size. The importance of preliminary screening of the prospective models is demonstrated in Problem 2.15, where inclusion of model f_4 yields considerably less impressive results.

The usefulness of DMs which directly contrast rival models and yield selection probabilities has been demonstrated. Procedures which measure separate model fittings alone, such as (2.6.1), do not do what is required in model discrimination, namely, to select the best model from a set of competitors and quantify that selection.

PROBLEMS

2.1 Two deterministic models f_1 and f_2 are compared where we define f^* as in Section 2.2.

(a) Show that a given by (2.2.1) minimizes SSR(f^*).

(b) Let $W = \sum_{j=1}^{n}(Y_j - f_j(1))(f_j(1) - f_j(2))$ and $V = \{\sum_{j=1}^{h}(Y_j - f_j(1))^2 \sum_{j=1}^{h}(f_j(1) - f_j(2))^2 - [\sum_{j=1}^{n}(y_j - f_j(1))(f_j(1) - f_j(2)]^2\}^{1/2}$. Show that [RSS($f_1$) – RSS($f^*$)]/ RSS($f^*$) is equivalent to $Q^2 = W/V$.

(c) Under model f_1, that is, $Y_j \sim$ i.i.d.$n(f_j(1), \sigma^2)$, what are the distributions of W and V^2?

(d) Based on (b) and (c) derive the distribution of (2.2.2).

2.2 Derive the Neyman–Pearson test of H_0: $E(Y_j) = f_j(1)$ for all $j \geq 1$ against H_1: $E(Y_j) = f_j(2)$ for all $j \geq 1$ under models (1.2.2) where, for true model f_i, $e_j(i) \sim$ i.i.d.$n(0, \sigma^2)$.

2.3 Consider the following problem, which consists of two models and data formed with two repetitions at each of five locations.

	Locations				
	0	.5	1.0	1.5	2.0
Y	−.02, .55	1.07, .05	1.19, 4.6	1.34, 1.8	2.06, 2.46
f_1	0	.5	1.0	1.5	2.0
f_2	0	.25	1.0	2.65	4.0

(a) Use (2.2.6) to calculate $\hat{\sigma}^2$.
(b) Which model is selected using LSE? Based on (a), estimate $P[\mathrm{IS}(f_i)]$ where f_i is selected.
(c) Based on the hypothesized true model (the one selected in (c)) estimate σ^2 by $\mathrm{RSS}(f_i)/n$. With this estimate calculate $P[\mathrm{IS}(f_i)]$ for the chosen model. How does this compare to the result in (b)?

2.4 For the models and data in Problem 2.3 we form a jackknife estimate of $P[\mathrm{IS}(f_i)]$. In sequence, we eliminate one data point and calculate $\delta(j)/s(j)$ for $j = 1, \ldots, 10$, where $s(j)$ is the value of $\hat{\sigma}$ and $\delta(j) = [\mathbf{f}(1,2)'\mathbf{f}(1,2)]^{1/2}$ with location $\mathbf{x}_j$ omitted. The jackknife estimate is $P_J^*[\mathrm{IS}(f_i)] = \sum_{j=1}^{n} P[Z > \delta(j)/(2s(j))]/n$. Compare this with Problem 2.3b. Does the jackknife estimate appear to reduce the bias present in estimating σ^2 (is this estimate raised)?

2.5 Consider the notation and setting of Section 2.3.

(a) Show that $\hat{\boldsymbol{\beta}}^*$ given by (2.3.1) minimizes RSS(f^*) subject to the constraint that the coefficients sum to one.

(b) If f^* is the true model, that is, Y is multivariate normal with mean vector $\mathbf{x}\boldsymbol{\beta}^*$ and dispersion matrix $\sigma^2\mathbf{I}$, what is the distribution of RSS(f^*)/σ^2?

(c) Under model correctness what is the distribution of RSS($\bar{f}$)/σ^2?

(d) If f is correct derive the distribution of (2.3.2).

2.6 We derive an alternative form of the Williams test by first defining the $k \times k$ matrix $\mathbf{V}$ consisting of elements $v_{hi} = \sum_{j=1}^{n}(Y_j - f_j(h))(Y_j - f_j(i))$.

(a) Show RSS(f) = $\mathbf{1}'\mathbf{V}\mathbf{1}/k^2$.

(b) Show $\hat{\boldsymbol{\beta}}^* = \mathbf{V}^{-1}\mathbf{1}/\mathbf{1}'\mathbf{V}^{-1}\mathbf{1}$ and RSS(f^*) = $1/\mathbf{1}'\mathbf{V}^{-1}\mathbf{1}$.

(c) How do we calculate (2.3.2)?

(d) The dispersion matrix of $\hat{\boldsymbol{\beta}}^*$ under correctness is $\sigma^2(\mathbf{V}^{-1} - \mathbf{1}'\mathbf{V}^{-1}\mathbf{1}\boldsymbol{\beta}^*\boldsymbol{\beta}^{*\prime})$. Construct a test of H_0: $\beta_1^* = \beta_2^*$ against H_1: $\beta_1^* \neq \beta_2^*$ where $\boldsymbol{\beta}^{*\prime} = (\beta_1^*, \ldots, \beta_k^*)$.

2.7 The example used by Williams concerned three models to determine the failing load of columns. The deterministic models were based on the slenderness of the column along with the strength of the material. The sums of squares and products of the residuals are given by

$$V = \begin{pmatrix} .00153 & .00436 & -.00005 \\ & .02072 & -.00459 \\ & & .00297 \end{pmatrix}$$

(see Problem 2.6).

(a) Calculate $\hat{\boldsymbol{\beta}}^*$.

(b) Run the Williams test at $\alpha = .05$.

(c) Use Problem 2.6d to compare the first and third models. Can we tell which is superior?

2.8 From the example introduced in Problem 2.7 we find the sums of squares and products of the model values to be

$$\mathbf{x}'\mathbf{x} = \begin{pmatrix} 1.9292 & 1.7719 & 2.0085 \\ & 1.6281 & 1.8438 \\ & & 2.0923 \end{pmatrix}$$

where the elements of $\mathbf{x}'\mathbf{x}$ are $\sum f_j(i)f_j(h)$ for $1 \leq i, h \leq 3$. Also $\sum y_j f_j(1) = 1.9585$, $\sum y_j f_j(2) = 1.7984$, and $\sum y_j f_j(3) = 2.0393$.

(a) Find $\mathbf{Z} = \mathbf{z}$ defined in the tilde test.

(b) Run the tilde test at $\alpha = .05$. Is there a difference between this test and the Williams test of Problem 2.7?

2.9 Apply the SM1 method to the Williams data using the sums of squares and products given in Problem 2.8. Let $f_{(1)} = f_2$, $f_{(2)} = f_1$, and $f_{(3)} = f_4$. Which model is selected? Estimate σ^2 based on the least favorable model (see Problem 2.3(c)) and evaluate the chosen model. Compare this result with that of Problem 2.7.

2.10 For the model values listed in Table 2.3 for Example 2.3.1 apply bound (2.3.11) under all chosen models. Compare these results with the simulation calculations in Table 2.7. Do you find the bounds to be useful?

2.11 The physicist in Example 2.2.1 considers an additional variable-acceleration model. Let f_3 correspond to the model whose height at time t is given by $h(t) = h(0) + .5gt^2 + ct^4$ where $c = 2$ ft/sec.

(a) Find the model function for f_3 where the time to hit the ground is recorded.
(b) Using the heights in Example 2.2.1, find the model function values for f_3.
(c) Since model f_1 is assumed correct under standard conditions, run the test outlined in (1.5.9) where f_1 is assumed correct under H_0 using LSE and (2.3.6) with $\alpha = .05$.
(d) Run the test of (c) over, using SM2 outlined in Table 2.6. Is this result better than for LSE?

2.12 Consider the SM2 procedure where $\mathbf{W}_1$ is a solution to $\mathbf{W}_1^* = \mathbf{G}_1\mathbf{G}_1'\mathbf{W}_1$.
(a) Let $\mathbf{Z}$ be another solution, that is, $\mathbf{Z}^* = \mathbf{G}_1\mathbf{G}_1'\mathbf{Z}_1$. Show $\mathbf{P}\mathbf{G}_1\mathbf{G}_1'\mathbf{W}_1 = \mathbf{G}_1'\mathbf{G}_1\mathbf{W}_1$ where $\mathbf{P}$ is diagonal and $\mathbf{Z} = \mathbf{P}\mathbf{W}$.
(b) When is $\mathbf{W}_1$ not unique?

2.13 In a Bayesian framework models have prior probabilities $P(f_i \text{ Correct}) = P(i)$ for $1 \le i \le k$. When $k \ge 3$ then (1.5.8) becomes $P(\text{IS}) \le \sum_{q \ne j} P(f_i \mid f_q)P(q)$. Consider this setting.
(a) Using LSE and (2.3.5) find bounds on $P(\text{IS})$.
(b) Rewrite Theorem 2.4.2 and, in particular, (2.4.22).

2.14 Three models computed at two locations, x_1 and x_2, are considered. We have $f_2(1) = f_1(2) = f_1(3) = f_2(3) = 1$ and $f_1(1) = f_2(2) = 2$. The following dynamic SM1 procedure is applied: use $f_{(1)} = f_1$ if f_1 is selected (using $f_{(1)} = f_1$) otherwise let $f_{(1)} = f_2$.
(a) Show that with $f_{(1)} = f_1$ or f_2 we find $d_1 = 1$, but with $f_1 = f_3$ we have $d_1 = 1/2^{1/2}$. Thus, the above procedure is valid.
(b) Let f_3 be the true model with errors $e_j(3)$ for $j = 1, 2$. Show that f_2 is selected if the event $\{e_1(3) < .5, e_2(3) > .5\}$ occurs.

(c) If $\sigma = 1$ find $P[\text{IS}(f_2 \mid f_3)]$ from (b).

(d) Suppose f_2 is selected and we estimate $P[\text{IS}(f_2)]$ by considering just the SM1 procedure with $f_{(1)} = f_2$ (consider $f_{(1)} = f_2$ fixed). Find $P[\text{IS}(f_2)]$ using $\sigma = 1$ and compare this value with that obtained in (c).

2.15 Consider the example in Section 2.6 with data listed in Table 2.8.

(a) Find $\hat{\sigma}^2$ using (2.2.6).

(b) Use LSE to select a best model.

(c) Use Theorem 2.3.1 to estimate $P[\text{IS}(f_i)]$ where f_i is chosen in (b). Compare this conditional probability to that for SM1 (.195).

2.16 Run the SM1 procedure on the data and model values given in Table 2.8. Let $f_{(1)} = f_4$, $f_{(2)} = f_3$, $f_3 = f_1$, and $f_{(4)} = f_2$ and estimate σ by using the method for repeat data points. Which model is selected? Compare $P[\text{IS}(f_i)]$ with the value found in Section 2.6, namely .195. Is there much advantage in eliminating f_4 before applying SM1?

3

Data-Fitted Models

3.1 INTRODUCTION

In this chapter we consider the most common setting for model discrimination, where the proposed model functions, sometimes referred to as the deterministic part of the modeling scheme, contain unknown parameters. These models, since they are estimated through parameter estimates based on the responses, are called data-fitted models. Unlike the completely determined models, which occur mainly in science or engineering contexts, these models are used in both the hard and soft sciences. Based on empirical data, the model's unknown parameters are estimated using the method of least squares. Substitution of the estimates for the models in S leads to k estimated or fitted models. In this setting we search for efficient DMs and DFs to apply to the set of competing models.

The fitted model function values of S take on a random nature since they depend on the data through their parameter estimates. For this reason, a proposed model is measured by two main characteristics. First, as in the case of completely determined models, is the fit of the estimated model function to the observed responses. This is assessed by DFs which are based on some function of the RSS. Second, parameter estimation causes a dependency between the estimated parameter and, hence, fitted model values and the particular observed data set. This characteristic we call model stability. The stability of a model may refer to either the variance of the estimated model function values or, if the parameters are of primary interest, the variance of the estimated parameters. A stable model with a moderate RSS may be preferable to an unstable model with a small RSS. For example, how useful is an mth degree polynomial model function fitted to n data responses in describing a future data set when m approaches n? An efficient model function should fit the responses to a desired accuracy while possessing a degree of stability over the range of application. For this reason we frequently use DMs which utilize one or more DFs that assess both fit and stability.

One of the most common DFs measured is the ratio of the variation accounted for by a proposed fitted model to the total variation in the responses. The total variation in Y_j, $1 \leq j \leq n$, or in the classical ANOVA sense the total sum of squares, is the result of applying the constant model function $\bar{Y}$ and is $\text{RSS}(\bar{Y}) = \sum_{j=1}^{n}(Y_j - \bar{Y})^2$. Further, if for $f_i \in S$, eliminating the i index, the fitted model function values are $\hat{f}_j$, $j \geq 1$, we note the partition $\text{RSS}(\bar{Y}) = \text{RSS}(\hat{f}_i) + \text{SSR}(\hat{f}_i)$ where $\text{SSR}(\hat{f}_i) = \sum(\hat{f}_j - \bar{Y})^2$ is referred to as the sum of squares due to the regression. A DF which assesses the fit of the proposed model function to the responses is the coefficient of variation, commonly called R^2, and is given by

$$R^2(\hat{f}_i) = 1 - \mathrm{RSS}(\hat{f}_i)/\mathrm{RSS}(\bar{Y}) \tag{3.1.1}$$

Here R^2 takes values between zero, indicating a random fit, and one, implying that the responses lie exactly on the fitted model function. This DF increases with the number of estimated parameters and is not sensitive to model stability. For this reason adjustments to (3.1.1) or other DFs are required. However, the change in R^2 has been proposed as a DF in nested model systems (Draper and Smith, 1981, pp. 296-297). Typical R^2 values change according to the area of research in which they are applied. In engineering problems an R^2 of .95 may be common, whereas in the social sciences an R^2 of .4 may be acceptable.

As mentioned in Section 1.3, the model function in the fitted model case is not unique in form. Any proposed model may be reparametrized for efficiency and utility. The effect of reparametrizing is discussed in Section 3.7, and for an overview we refer to Ratkowsky (1983). As in Bates and Watts (1980), we take the term model function to mean a designation of a particular parametrization. Hence, two different reparametrizations are treated as two competing models since they have different stability characteristics in terms of estimated parameters and truncation properties.

With estimated models it often happens that Atkinson's question A2 (under the assumption that the models fit the data equally well, is there statistical evidence to the contrary?) is asked. This is the case in data exploration analysis as used, for example, in the social sciences. Here, DFs which assess fit and stability at individual locations are constructed and used together in a DM. In some cases, such as linear regression modeling, a DM may consist of only one DF which accounts for both fit and stability. Under A2, since a true model is not assumed, selection probabilities do not come into play.

For model discrimination problems where the model functions are constructed on the basis of physical properties and laws, questions A1 and A3, to a limited degree, may be considered. In some cases, as in the completely determined model setting, selection probabilities can be calculated or estimated and used as the basis for DMs.

The advent of computer simulation and computations has allowed new analytic procedures. These methods open up further analysis along the lines of Atkinson's questions. While Chapter 4 takes up these methods in detail, this chapter lays the groundwork for the basic DFs to be utilized.

We conclude this section with an analog of the preliminary test (2.1.1) in the estimated model case. When repeat data points are present, the estimator of σ^2 independent of the fitted model functions, namely $\hat{\sigma}^2$ given by (2.2.6), is used. In this context m denotes the degrees of freedom associated with $\hat{\sigma}^2$. For $f_i \in \mathcal{S}$ the mean squares lack of fit is $\mathrm{MSL}(\hat{f}_i) = [\mathrm{RSS}(\hat{f}_i) - \hat{\sigma}^2 m]/[n - m(i) - m]$. The resulting lack-of-fit statistic for f_i is

$$\mathrm{LOF}(\hat{f}_i) = P[F > \mathrm{msl}(\hat{f}_i)/(\hat{\sigma}^2/m)] \tag{3.1.2}$$

where $\mathrm{msl}(\hat{f}_i)$ is the observed value of MSL and the random variable F has $n - m(i) - m$ and m degrees of freedom. If the model function is linear in parameters then LOF is exact. On the other hand, if the model function is nonlinear then (3.1.2) is an approximation and the degree of its accuracy depends on the ability of the estimates of the model function values to be linearized. This topic is discussed in Section 3.7. When applicable, the DF defined in (3.1.2) provides a means of eliminating unlikely models.

3.2 ESTIMATION

The investigation into the model discrimination problem, under any of Atkinson's questions, starts with an examination of the fitted model function's structure. The asymptotic theory of the parameter estimates, in the context of nonidentical sampling location, is an extension of the general theory (Malinvaud, 1970, Chapter 9). Based on these asymptotic forms, the DFs which constitute the proposed DMs in the estimated model case are constructed.

Consider an arbitrary model from S consisting of p predictor or independent variables. As outlined in Section 1.2, the responses Y_j are associated with the locations $\mathbf{x}_j' = (x_{j1},\ldots,x_{jp})$ for $1 \leq j \leq n$. The model function is a function of both $\mathbf{x}_j$ and m unknown parameters given in $\boldsymbol{\theta}' = (\theta_1,\ldots,\theta_m)$. Hence, eliminating the i index, the model function values at $\mathbf{x}_j$ are $f_j(\boldsymbol{\theta})$ for $j \geq 1$. The $n \times 1$ vector of model function values is denoted $\mathbf{f}(\boldsymbol{\theta})$. The modeling scheme similar to (1.2.2) is

$$Y_j = f_j(\boldsymbol{\theta}) + e_j \tag{3.2.1}$$

for $j \geq 1$ and $1 \leq i \leq k$. As in the previous chapters, normality of the error term is assumed, namely, $Y_j \sim \text{i.i.d.}n(\mu_j, \sigma^2)$ for $j \geq 1$. A model from S is admissible if there exists a $\boldsymbol{\theta}_0' = (\theta_{01},\ldots,\theta_{0m})$ such that $\mu_j = f_j(\boldsymbol{\theta}_0)$ for $j \geq 1$. A model is linear in parameters, or simply linear, if $(\partial/\partial\theta_t)f_j(\boldsymbol{\theta}) = c_j$ for $1 \leq j \leq n$ and $1 \leq t \leq m$.

To estimate the unknown parameters in the models we use the method of least squares, denoted LS. For any proposed model we select to minimize

$$\text{RSS}(f_i) = \sum_{j=1}^{n} [y_j - f_j(\boldsymbol{\theta})]^2 \tag{3.2.2}$$

The resulting LS estimator is denoted $\hat{\boldsymbol{\theta}}' = (\hat{\theta}_1, \ldots, \hat{\theta}_m)$. The fitted or estimated model function values are obtained by substitution of the LS estimator and are denoted $\hat{f}_j (= f_j(\hat{\boldsymbol{\theta}}))$. The LS estimator $\hat{\boldsymbol{\theta}}$ is assumed to be well defined. The first theorem dealing with the properties of θ under normality of the errors is well known and is given for completeness.

Theorem 3.2.1

Fix a model in S where (3.2.1) holds and $e_j \sim$ i.i.d.$n(0, \sigma^2)$ for $j \geq 1$ and $\boldsymbol{\theta} = \boldsymbol{\theta}_0$. The LS estimator $\hat{\boldsymbol{\theta}}$, assuming it exists, coincides with the maximum likelihood estimator.

Proof Using the normality of the error term, the log likelihood function for the data is

$$\ln[L(\boldsymbol{\theta})] = -n \ln(2\pi)/2 - n \ln(\sigma^2)/2 - \sum_{j=1}^{n} [y_j - f_j(\boldsymbol{\theta})]^2/(2\sigma^2)$$

Taking the derivative with respect to σ^2 and θ_t, $1 \leq t \leq m$, yields maximum likelihood estimates $\hat{\boldsymbol{\theta}}$ and $\hat{\sigma}_f^2 = \sum_{j=1}^{n} [Y_j - f_j(\hat{\boldsymbol{\theta}})]^2/n$.

To solve for the LS estimator based on maximizing (3.2.2) many methods have been proposed. Before discussing one such method the large sample properties of $\hat{\boldsymbol{\theta}}$ are examined. To do this we first introduce some notation and terminology proposed by Jennrich (1969). Let two sequence-valued functions (functions of a location index) defined on R^m be given by $g_j(\boldsymbol{\theta})$ and $h_j(\boldsymbol{\theta})$. For $\boldsymbol{\alpha},\boldsymbol{\beta} \in R^m$ define

$$(g_j(\boldsymbol{\alpha}),h_j(\boldsymbol{\beta}))_n = \sum_{j=1}^{n} g_j(\boldsymbol{\alpha})h_j(\boldsymbol{\beta})/n \tag{3.2.3}$$

If $g_j(\boldsymbol{\alpha}) = h_j(\boldsymbol{\alpha})$ we denote (3.2.3) by $(g_j(\boldsymbol{\alpha}))_n^2$. The tail cross product of $g_j(\boldsymbol{\alpha})$ and $h_j(\boldsymbol{\beta})$ exists if (3.2.3) converges uniformly in $\boldsymbol{\alpha}$ and $\boldsymbol{\beta}$, that is,

$$(g_j(\boldsymbol{\alpha}),h_j(\boldsymbol{\beta}))_n \longrightarrow [g(\boldsymbol{\alpha}),h(\boldsymbol{\beta})] \tag{3.2.4}$$

uniformly in $\alpha,\beta \in R^m$. The right-hand side of (3.2.4) is the limit function and when $g_j(\boldsymbol{\alpha}) = h_j(\boldsymbol{\alpha})$ is denoted $[g](\alpha)$.

Fixing a model from $\mathcal{S}$, the first- and second-order partial derivatives, assuming they exist, are denoted by $f_{jr}(\boldsymbol{\theta}) = (\partial/\partial\theta_r) \times f_j(\boldsymbol{\theta})$ and $f_{jrs}(\boldsymbol{\theta}) = (\partial^2/\partial\theta_r\partial\theta_s)f_j(\boldsymbol{\theta})$. Let $a_{rs}(\boldsymbol{\theta})_n = (f_{jr}(\boldsymbol{\theta}),f_{js}(\boldsymbol{\theta}))_n$. Further, at location $\mathbf{x}_j$ define the vector of first-order partial derivatives by $\mathbf{F}_j(\boldsymbol{\theta}) = (f_{j1}(\boldsymbol{\theta}),\ldots,f_{jm}(\boldsymbol{\theta}))$ and the matrix of second-order partial derivatives by $\mathbf{H}_j(\boldsymbol{\theta}) = ((f_{jrs}(\boldsymbol{\theta})))$ for $1 \le j \le n$. Also, define the $m \times n$ matrix of derivatives $\mathbf{F}(\boldsymbol{\theta})' = (\mathbf{F}_1(\boldsymbol{\theta}),\ldots,\mathbf{F}_n(\boldsymbol{\theta}))$. Further, let $\mathbf{A}(\boldsymbol{\theta})_n$ be the $m \times m$ matrix consisting of elements $a_{rs}(\boldsymbol{\theta})_n$. In keeping with the notation for the fitted model values, let $\hat{\mathbf{F}}_j, \hat{\mathbf{H}}_j, \hat{\mathbf{F}}, \hat{a}_{rs}$ and $\hat{\mathbf{A}}$ be the appropriate functions evaluated at $\hat{\boldsymbol{\theta}}$.

Two theorems due to Jennrich concerning the asymptotic consistency and normality of the LS estimator follow. The proofs are omitted and we refer to their original source.

Theorem 3.2.2

Fix a model from $\mathcal{S}$ and let $f_j(\boldsymbol{\theta})$ be continuous in $\boldsymbol{\theta}, \boldsymbol{\theta} \in \Omega$, a compact subset of R^m. Let (3.2.1) hold where $\boldsymbol{\theta} = \boldsymbol{\theta}_0$ implies that f_i is admissible and e_j are independent with $E[e_j] = 0$ and $E[e_j^2] = \sigma^2$ for $j \geq 1$. Let $[f](\boldsymbol{\theta})$ exist in the sense of (3.2.4) and assume $[f(\boldsymbol{\theta}), f(\boldsymbol{\theta}_0)]$ possess a unique minimum at $\boldsymbol{\theta} = \boldsymbol{\theta}_0$. Then $\hat{\boldsymbol{\theta}} \to \boldsymbol{\theta}_0$ and $\hat{\sigma}_f^2 \to \sigma^2$ with probability one as $n \to \infty$.

Theorem 3.2.3

Assume the conditions of Theorem 3.2.2 and let $f_{jr}(\boldsymbol{\theta})$ and $f_{jrs}(\boldsymbol{\theta})$ exist and be continuous in $\boldsymbol{\theta}$. Let $[h,g](\boldsymbol{\theta})$ exist where $h_j(\boldsymbol{\theta})$, $g_j(\boldsymbol{\theta}) = f_j(\boldsymbol{\theta}), f_{jr}(\boldsymbol{\theta})$ and $f_{jrs}(\boldsymbol{\theta})$. Then $a_{rs}(\boldsymbol{\theta})_n$ converges to constants $a_{rs}(\boldsymbol{\theta})$ as n increases. Define the $m \times m$ matrix with elements $a_{rs}(\boldsymbol{\theta})$ by $\mathbf{A}(\boldsymbol{\theta})$. If $\boldsymbol{\theta}_0$ is an interior point of Ω and $\mathbf{A} = \mathbf{A}(\boldsymbol{\theta}_0)$, consisting of elements a_{rs}, is nonsingular, then

$$L\{n^{1/2}(\hat{\boldsymbol{\theta}} - \boldsymbol{\theta}_0)\} \longrightarrow N(0, \sigma^2\mathbf{A}^{-1}) \tag{3.2.5}$$

as $n \to \infty$ where $\hat{a}_{rsn} \to a_{rs}$ with probability one.

The assumptions put forth by Jennrich in Theorems 3.2.2 and 3.2.3 are referred to as non-linear regularity conditions or NRCs. Other work on the asymptotic properties of the LS estimator has appeared in the literature, and we refer to Chien (1981), Drygas (1976), and Lai et al. (1978). These concern extensions of the classical large sample theory (see Wald, 1949, and Cramer, 1946).

In general it is not easy to present practical guidelines for the convergence of $\hat{\boldsymbol{\theta}}$ which can be verified analytically by the experimenter. The NRCs concerning differentiability are easily checked, but convergence in the sense of (3.2.4), even though of mathematical importance, is another matter. Simulation procedures can be used to check the accuracy of these approximations (see Chapter 4). As long as the predictor variables take values in a bounded region in which the model functions are continuously differentiable, the NRCs can reasonably be assumed to hold. The lemma which follows is used to construct a large sample estimator of the asymptotic dispersion matrix $\mathbf{A}^{-1}$ in (3.2.5).

Lemma 3.2.1

Let the sequence-valued function $h_j(\boldsymbol{\alpha})$ and $g_j(\boldsymbol{\alpha})$ be continuous in $\boldsymbol{\alpha}$ and assume $[g(\boldsymbol{\alpha}),h(\boldsymbol{\beta})]$ exists for $\boldsymbol{\alpha},\boldsymbol{\beta} \in R^m$. If $\boldsymbol{\alpha} \to \boldsymbol{\alpha}_0$ and $\boldsymbol{\beta} \to \boldsymbol{\alpha}_0$ with probability one as $n \to \infty$ then $(g_j(\boldsymbol{\alpha}),h_j(\boldsymbol{\beta}))_n \to [g,h](\alpha_0)$ with unit probability as $n \to \infty$.

Proof Continuity of the functions implies that $[g(\boldsymbol{\alpha}),h(\boldsymbol{\beta})]$ is continuous. Fix $\epsilon > 0$ and notice that there exists a $\delta > 0$ such that if $\|\boldsymbol{\alpha},\boldsymbol{\alpha}_0\|_\infty < \delta$ and $\|\boldsymbol{\beta},\boldsymbol{\beta}_0\|_\infty < \delta$ then $|[g,h](\boldsymbol{\alpha}_0) - [g(\boldsymbol{\alpha}),h(\boldsymbol{\beta})]| < \epsilon/2$. Since $[g(\boldsymbol{\alpha}),h(\boldsymbol{\beta})]$ exists there is an $N_1 > 0$ such that for $n \geq N_1$, $\mathrm{Sup}|(g_j(\boldsymbol{\alpha}),h_j(\boldsymbol{\beta}))_n - [g(\boldsymbol{\alpha}),h(\boldsymbol{\beta})]| < \epsilon/2$. Strong consistency of $\boldsymbol{\alpha}$ and $\boldsymbol{\beta}$ imply, with probability one, that there exists an $N_2 > 0$ such that for $n \geq N_2$, $\|\boldsymbol{\alpha},\boldsymbol{\alpha}_0\|_\infty < \delta$ and $\|\boldsymbol{\beta},\boldsymbol{\beta}_0\|_\infty < \delta$. Hence, for $n \geq \max\{N_1,N_2\}$, $|(g_j(\boldsymbol{\alpha}),h_j(\boldsymbol{\beta}))_n - [g,h](\boldsymbol{\alpha}_0)| < \epsilon$ with probability one. Noting that ϵ is arbitrary finishes the proof.

For the true model, under NRC, Lemma 3.2.1 and the previous theorems imply, with probability one, $\hat{a}_{rsn} = (f_{jr}(\hat{\boldsymbol{\theta}}),f_{js}(\hat{\boldsymbol{\theta}}))_n \to a_{rs}$ as $n \to \infty$. Thus $\sum \hat{\mathbf{F}}_j \hat{\mathbf{F}}_j'/n = F(\hat{\boldsymbol{\theta}})'F(\hat{\boldsymbol{\theta}})/\mathbf{n} \to \mathbf{A}$, and defining $M(\boldsymbol{\theta}) = (F(\boldsymbol{\theta}),F(\boldsymbol{\theta})')^{-1}$ we note that $n\hat{M} = \mathbf{nM}(\hat{\boldsymbol{\theta}}) \to A^{-1}$ with unit probability as $n \to \infty$. Hence, under NRC, when considering thc true model, $\hat{\boldsymbol{\theta}}$ is distributed asymptotically normal with mean $\boldsymbol{\theta}_0$ and approximate dispersion matrix $\sigma^2\hat{\mathbf{M}}$. In the classical asymptotic theory $F(\boldsymbol{\theta})F(\boldsymbol{\theta})'/\sigma^2$ is the information matrix.

A great many iterative procedures have been proposed for finding the LS estimator for nonlinear models (see review paper by Chambers, 1973). We focus on one method that possesses quick convergence characteristics for models that are "close" to linearity. This topic is discussed in Section 3.7. The modified Gauss–Newton procedure for finding $\hat{\boldsymbol{\theta}}$ employs the iterative system outlined by

$$\boldsymbol{\theta}_{r+1} = \boldsymbol{\theta}_r + \mathbf{M}(\boldsymbol{\theta}_r)[\mathbf{F}(\boldsymbol{\theta}_r)'\mathbf{Y} - f(\boldsymbol{\theta}_r)] \tag{3.2.6}$$

for $r \geq 1$. Applying an initial estimate $\boldsymbol{\theta}$ leads to the series $\boldsymbol{\theta}_1, \boldsymbol{\theta}_2, \ldots$. We stop and declare convergence when $(\boldsymbol{\theta}_{r+1} - \boldsymbol{\theta}_r)'(\boldsymbol{\theta}_{r+1} - \boldsymbol{\theta}_r)$ is small. The properties of this iterative method will not be explored here but we refer to Ratkowsky (1983, p. 155) for a discussion. To form DMs and DFs we utilize the large sample characteristics of the LS estimator.

The asymptotic theory of $\hat{\boldsymbol{\theta}}$ is extended to the fitted model function values by a construction due to Box (1971). Under the true model the "correct" model function (for $\boldsymbol{\theta} = \boldsymbol{\theta}_0$) values are $f_j = f_j(\boldsymbol{\theta}_0)$ for $1 \leq j \leq n$. The estimated model function values are denoted $\hat{f}_j = f_j(\hat{\boldsymbol{\theta}})$ for $j \geq 1$. Using a Taylor series expansion for $f_i \in \mathcal{S}$, we obtain the approximate model function values

$$\hat{f}_j = f_j + \mathbf{F}_j'\boldsymbol{\phi} + \boldsymbol{\phi}'\mathbf{H}_j(\boldsymbol{\theta}^*)\boldsymbol{\phi}/2 \tag{3.2.7}$$

where $\boldsymbol{\phi} = \hat{\boldsymbol{\theta}} - \boldsymbol{\theta}_0$ and $\boldsymbol{\theta}^*$ is a suitable vector in R^m (see Apostel, 1975, p. 361) and $\mathbf{F}_j = \mathbf{F}_j(\boldsymbol{\theta}_0)$. The following theorem is due to Box.

Theorem 3.2.4

For a true model with model function (3.2.1) assume that the relation between $\boldsymbol{\phi}$ and the vector of errors, denoted by $\mathbf{e}' = (e_1, \ldots, e_n)$,

$$\boldsymbol{\phi} = \mathbf{De} + \mathbf{q} \tag{3.2.8}$$

holds where $\mathbf{D}$ is $m \times n$ and $\mathbf{q}' = (\mathbf{e}'\mathbf{B}_1\mathbf{e}, \ldots, \mathbf{e}'\mathbf{B}_m\mathbf{e})$ for constant $n \times n$ matrices $\mathbf{B}_r$ for $1 \leq r \leq m$. Since $\hat{\boldsymbol{\theta}}$ is an LS estimator, $\mathbf{D} = \mathbf{F}$. also, as $n \to \infty$

$$E(\phi) = -\sigma^2\mathbf{MF}'\mathbf{G}/2 \tag{3.2.9}$$

where $\mathbf{G}$ is an $n \times 1$ vector composed of elements $\mathrm{Tr}[\mathbf{H}_j\mathbf{M}]$ and $\mathbf{H}_j = \mathbf{H}_j(\boldsymbol{\theta}_0)$ for $j \geq 1$.

Proof Taking the derivative of (3.2.2), we find that $\hat{\boldsymbol{\theta}}$ satisfies $\mathbf{F}'[\mathbf{Y} - \mathbf{f}] = \mathbf{0}$. Using approximation (3.2.7), this becomes $\mathbf{F}'[\mathbf{Y} - \mathbf{F}\phi - \mathbf{J}] = \mathbf{0}$ where $\mathbf{J}' = (\phi'\mathbf{H}_j(\boldsymbol{\theta}^*)\phi/2, \ldots, \phi'\mathbf{H}_n(\boldsymbol{\theta}^*)\phi/2)$. Putting (3.2.8) into this relation yields $\mathbf{D} = \mathbf{F}$. also, (3.2.9) is obtained by taking the expected value of the previous form.

Theorem 3.2.4 extends the asymptotic properties of given in Theorem 3.2.2. The large sample normality in connection with (3.2.8) allows us to drop the quadratic function of the errors, $\mathbf{q}$, in forming the approximation

$$\phi = \mathbf{MF}'\mathbf{e} + o(n^{-1/2}) \tag{3.2.10}$$

under the true model. This combination of the works of Jennrich and Box plays a central role in the development of asymptotic DFs. Further, the well-known property that nonlinear models lead to biased LS estimators is apparent from (3.2.9). Under NRC, combining (3.2.7), (3.2.9), and (3.2.10), the amount of bias in the estimated model function at location $\mathbf{x}_j$ is

$$\beta_j = \sigma^2\,\mathrm{Tr}[\mathbf{H}_j\mathbf{M}]/2 - \mathbf{F}_j'\mathbf{MF}'\mathbf{G} + o(n^{-1/2}) \tag{3.2.11}$$

Other approximations based on higher-order estimates are possible and we refer to Clarke (1980). From convergence properties an estimate of β_j is

$$\hat{\beta}_j = \hat{\sigma}^2 \,\mathrm{Tr}[\hat{\mathbf{H}}_j \hat{\mathbf{M}}]/2 - \hat{\mathbf{F}}_j' \hat{\mathbf{M}} \hat{\mathbf{F}} \hat{\mathbf{G}} \qquad (3.2.12)$$

where $\hat{\sigma}^2$ is an estimate of σ^2, such as (2.2.6).

The asymptotic theory presented here is used as a guide in the construction of DFs and DMs. The correctness of these approximations in specific, possible small sample size, cases can be evaluated by utilizing simulation procedures.

We now follow with a theorem that measures the effect of nonlinearity and model correctness on RSS or (3.2.2). Suppose model f_2 is proposed for the responses when, in fact, model f_1 is correct (for $\boldsymbol{\theta} = \boldsymbol{\theta}_0$). The model functions are denoted $f_j(i, \boldsymbol{\theta})$ for $i = 1,2$ and the LS estimators for both models are found. In the notations of this section the estimated model function values are $\hat{f}_j(i)$ while the limit model function values are $f_j(i)$ obtained substituting $\boldsymbol{\theta}_0$ and α. Also, let $\delta_n = (f_j(1) - f_j(2))_n^2$.

Theorem 3.2.5

Let f_1 be correct, let NRC hold, and let $[f(1, \boldsymbol{\theta}_0), f(2, \boldsymbol{\alpha})]$ exist in the sense of (3.2.4) with a unique minimum at $\boldsymbol{\alpha}_0$. If $\delta_n/n \to 0$ as $n \to \infty$ then

$$\mathrm{MRSS}(\hat{f}_2) = \sigma^2 + \delta_n - \mathrm{Var}(\hat{f}_2)_n + 2\,\mathrm{Cov}(Y - \hat{f}_2, \hat{f}_2)_n + o(1) \quad (3.2.13)$$

with unit probability, where $\mathrm{Var}(\hat{f}_2)_n = (\hat{f}_j(2) - f_j(2))_n^2$, $\mathrm{Cov}(Y - \hat{f}_2, \hat{f}_2)_n = (Y_j - \hat{f}_j(2), \hat{f}_j(2) - f_j(2))_n$, and MRSS is the mean residual sum of squares, or RSS/n.

Proof The proof is left as an exercise.

From Theorem 3.2.5 we note that an LQM based on RSS is valid only when comparing models with equivalent values of $\mathrm{Var}(\hat{f})_n$ and $\mathrm{Cov}(Y - \hat{f}, \hat{f})_n$. These quantities involve model function stabilities and nonlinearities (Problem 3.3). A positive δ_n, implying an incorrect model, may be masked in terms of RSS or MRSS by a positive $\mathrm{Var}(\hat{f})_n - \mathrm{Cov}(Y - \hat{f}, \hat{f})_n$ value. For this reason DFs which assess both model fit and stability, which is tied to model nonlinearity, are needed. Recently, a diagnostic tool which accounts for model nonlinearities called projected residuals has been proposed by Cook and Tsai (1985).

Remark 3.2.1

An alternative estimate of the asymptotic dispersion matrix of $\hat{\boldsymbol{\theta}}$ follows from the normal equations $\mathbf{F}'(Y - \mathbf{f}) = \mathbf{0}$. Expanding both F and **f** in Taylor series and discarding quadratic and higher-order terms in $\boldsymbol{\phi} = \hat{\boldsymbol{\theta}} - \boldsymbol{\theta}$ yields

$$\mathbf{F}'\mathbf{F}\boldsymbol{\phi} + \mathbf{F}'\mathbf{e} + \mathbf{G}'\mathbf{e} = \mathbf{0}$$

where **G** is the $n \times m$ matrix consisting of elements $\sum_t f_{jrt} \phi_t$ for $1 \leq j \leq n$ and $1 \leq r \leq m$. Showing $\mathbf{G}'\mathbf{e} = \mathbf{W}\boldsymbol{\phi}$, where **W** consists of elements $\sum_j f_{jrt} e_j$, we find the large sample approximation

$$\boldsymbol{\phi} = \mathbf{T}^{-1}\mathbf{F}'\mathbf{e} \qquad (3.2.14)$$

where $\mathbf{T} = \mathbf{F}'\mathbf{F} - \mathbf{W}$. Here $\hat{\mathbf{T}}$ is the observed information matrix, which, as suggested by (3.2.14), can be used in place of **M** in asymptotic statistics.

The modeler can use either $\hat{\mathbf{M}}$ of $\hat{\mathbf{T}}^{-1}$ in the DFs which follow, but for consistency we use $\hat{\mathbf{M}}$ in the following formulas. It has been suggested that the observed information matrix is more accurate when estimating the dispersion matrix (see Example 4.3.1).

3.3 NESTED MODELS

The modeler of a physical or natural system often starts with a basic model. More complexity, along with an increasing number of unknown parameters, is continually added, thereby forming a sequence of proposed models. With increasing complexity the model functions become morc accurate in their fit to the responses but their instability grows. In this way a natural hierarchical system of model functions is constructed. This embedding of model functions is called a nested system. Atkinson's question A1, assuming the most basic model is correct, is utilized to test "up" the chain of competing model functions.

Let $\mathcal{S}$ consist of a completely nested system of models given by $f_1 \subset f_2 \subset \cdots \subset f_k$ where $f_i \subset f_r$ implies that model function $f_j(i, \boldsymbol{\theta})$ is a reduction of $f_j(r, \boldsymbol{\theta})$ (see Example 1.3.1). For $f_i \in \mathcal{S}$ the model function contains unknown parameter vector $\boldsymbol{\theta}(i)$, which is $1 \times m(i)$. Because of the nesting $m(1) < m(2) < \cdots < m(k)$ and for $i < t$ there exists a $\boldsymbol{\theta}(t)^* = \boldsymbol{\theta}(t)$ (with some components zero) such that $f_j(i, \boldsymbol{\theta}(i)) = f_j(t, \boldsymbol{\theta}(t)^*)$ for $j \geq 1$. Also, as more complexity is added in the form of more parameters we have $\text{RSS}(f_1) \geq \text{RSS}(f_2) \geq \cdots \geq \text{RSS}(f_k)$. A general procedure for testing nested models in the arbitrary model function setting is now given.

We compare two models using the likelihood ratio test procedure. For $f_i \subset f_q$ we test

$$H_0\colon E(Y_j) = f_j(i, \boldsymbol{\theta}(i)), \boldsymbol{\theta}(i) \in R^{m(i)}, \qquad j \geq 1$$

against

$$H_1\colon E(Y_j) = f_j(q, \boldsymbol{\theta}(q)), \boldsymbol{\theta}(q) \in R^{m(q)}, \qquad j \geq 1 \tag{3.3.1}$$

Under correct model f_r with modeling scheme (3.2.1) and normality of the error we have $\mathbf{Y} \sim N(f(r, \boldsymbol{\theta}(r)), \sigma^2\mathbf{I})$ for $r = i$ or q. For the likelihood ratio test we find Sup($\mathbf{Y}$) or

$$\mathrm{Sup}(2\pi\sigma^2)^{-n/2} \exp\left\{-\sum[Y_j - f_j(r, \boldsymbol{\theta}(r))]^2/(2\sigma^2)\right\} \tag{3.3.2}$$

With respect to $\boldsymbol{\theta}(r)$ and σ^2, derivatives yield the LS estimator $\boldsymbol{\theta}(r)$ and $\hat{\sigma}_f^2 = \sum[Y_j - f_j(r, \boldsymbol{\theta}(r))]^2/n$ for $f = f_i$ or f_q. Under the specific models of hypotheses (3.3.1), (3.3.2) becomes, upon substitution of estimates $\hat{\boldsymbol{\theta}}(r)$, $(2\hat{\sigma}_{fr}^2)^{-n/2}\exp(-n/2)$ where under $H_0 r = i$ and under $H_1 r = q$. The likelihood ratio statistic is $\lambda = \lambda_i/\lambda_q$, which simplifies to

$$\lambda = [\mathrm{RSS}(\hat{f}_q)/\mathrm{RSS}(\hat{f}_i)]^{n/2} \tag{3.3.3}$$

We reject h_0 if λ is small. Under general regularity conditions, similar to NRC, $-2\ln(\lambda) \sim \chi^2$ with degrees of freedom $m(q) - m(i)$. Hence, an α significance level critical region is $n \ln[\mathrm{RSS}(f_i)/\mathrm{RSS}(f_q)] > \chi^2_{(1-\alpha)}$.

The test of (3.3.1) outlined above is referred to as the generalized likelihood ratio test (GLRT) for nested models. It is appropriate for both linear and nonlinear model functions. If the tested models have linear model functions, then an exact test based on (3.3.3) exists. To see this we write $\lambda^{2/n}$ as $(1 - \lambda^*)/\lambda^*$ where $\lambda^* = [\mathrm{RSS}(\hat{f}_i) - \mathrm{RSS}(\hat{f}_q)]/\mathrm{RSS}(\hat{f}_q)$. On finding the distribution of λ^* under normality of the error and true model f_i, we derive the test statistic for the linear model case. This test statistic is given by (1.5.10) where, since one of the predictor variables (associated with the intercept) is identically one, $r_i + 1$ and $r_q + 1$ are replaced by $m(i)$ and $m(q)$, respectively.

A common application of testing nested models is to test for what is called the significance of the total regression. To do this the constant model function $\bar{Y}$, taking the value $\bar{y}$ at each location, is assumed to be contained in the proposed model function. Based on the GLRT of $\bar{Y}$ against $f_i \in \mathcal{S}$, a DF for the significance of the regression of f_i is

$$\mathrm{SGF}(\hat{f}_i) = P[\chi^2_{(m(i)-1)} > n \ln[\mathrm{rss}(\bar{Y})/\mathrm{rss}(\hat{f}_i)]] \tag{3.3.4}$$

where rss is the observed value of RSS. If model function $f_j(i, \boldsymbol{\theta}(i))$ is linear in parameters using (1.5.10), the standard test of significance, contrasting $\bar{Y}$ and f_i, is obtained. Here the residual mean square based on f_i, given by

$$S^2(\hat{f}_i) = \mathrm{RSS}(\hat{f}_i)/[n - m(i)] \tag{3.3.5}$$

is utilized. The DF for significance of the regression in this case is

$$\mathrm{SGF}_L(\hat{f}_i) = P\left\{F > [\mathrm{rss}(\bar{Y}) - \mathrm{rss}(\hat{f}_i)]/[(m(i)-1)s^2(\hat{f}_i)]\right\} \tag{3.3.6}$$

where $F \sim F_{(m(i)-1, n-m(i))}$ and s^2 is the observed value of (3.3.5). These DFs are simply the observed significance levels or p-values for the corresponding nested model's tests.

Example 3.3.1

Compartmental models measuring the concentration of a substance at time t in a component system are contrasted. These models, developed by Bailey et al. (1976), are applied to the data of Gale and Eadie. The models monitor the progress of hydrolysis of urea through the rate of evolution of CO_2, denoted $C(t)$. The model functions are f_1: $C(t) = 100[1 - \exp(-\theta_1 t)(1 + \theta_1 t)]$, f_2: $C(t) = 100[1 - \theta_2 \exp(-\theta_1 t)/(\theta_2 - \theta_1) + \theta_1 \exp(-\theta_2 t)/(\theta_2 - \theta_1)$, and f_3: $C(t) = 100[1 - \theta_2 \exp(-\theta_1(t + \theta_3))/(\theta_2 - \theta_1) + \theta_1 \exp(-\theta_2(t + \theta_3))/(\theta_2 - \theta_1)]$. Here we note only f_2 is a nested model with $f_2 \subset f_3$. At specific times the observed percentage of CO_2 is recorded and the LS estimators for the model functions are found. These data are listed in Table 3.1. A dramatic drop in RSS is observed as the model functions become more complex utilizing additional parameters. Because of the large RSS for f_1 it is eliminated from consideration. It is interesting to note that R^2, given by (3.1.1), is not helpful in this case since $R^2(f_1) = .988$ and $R^2(f_i) \geq .99$ for $i = 2$ and 3. Repeat data points would have been helpful, as a test of model adequacy such as (2.1.1) could have been applied. The GLRT test based on (3.3.3) is run, yielding $-2\ln\lambda = 19.03$. Noting $\chi^2_{(.99,1)} = 6.63$, we reject H_0, given by (3.3.1), with $i = 2$ and $q = 3$, at significance level $\alpha = .01$. A significant gain in fit to the responses is obtained by the complexity added in going from f_2 to f_3.

Remark 3.3.1

In some cases the exact linear test given by (1.5.10) has been suggested as an approximate procedure in the nonlinear setting. Pettersson and Pettersson (1970) applied the F test to a set of compartmental models for enzyme kinetic data. The power of such a test was explored by Burguillo et al. (1983).

A model function's utility is in part a function of its predictability on future data sets. A method developed by Geisser (1975), called the predictive sample reuse method, or PSR, attempts to assess a model's usefulness on new data without taking additional

Table 3.1 Compartmental Models—Percentage of CO_2

Time	Responses	f_i	f_2	f_3
2	5.68	5.68	7.77	9.39
4	22.4	17.8	21.8	22.2
6	35.6	31.8	35.8	35.4
8	47.0	45.2	47.9	47.3
10	57.6	56.9	58.0	57.5
12	65.9	66.7	66.2	66.0
14	72.0	74.7	72.8	72.9
16	78.6	80.9	78.2	78.5
18	83.4	85.7	82.5	82.9
20	86.1	89.4	85.6	86.4
RSS		77.4	4.66	.695
$\hat{\boldsymbol{\theta}}'$		.191	.1098	.1185
$\hat{\boldsymbol{\theta}}'$			.5283	.3355
$\hat{\boldsymbol{\theta}}'$				.6948

samples. This procedure is based on the GLRT. For any model from $\mathcal{S}$ let $\hat{\boldsymbol{\theta}}_{(j)}$ be the LS estimate obtained by eliminating one data point, namely $(Y_j, \mathbf{x}_j)$, for $1 \leq j \leq k$. For a proposed model the predictive density at $\mathbf{x}_j$ is defined as the probability density function corresponding to Y_j evaluated at $\hat{\boldsymbol{\theta}}_{(j)}$. This function is denoted $d_j(\hat{\boldsymbol{\theta}}_{(j)})$. Under normality of the error term in (3.2.1), $d_j(\hat{\boldsymbol{\theta}}_{(j)}) = (2\pi\sigma^2)^{-n/2}\exp\left\{-\sum[Y_j - f_j(\hat{\boldsymbol{\theta}}_{(j)})]^2/(2\sigma^2)\right\}$. The value of σ^2 corresponding to the supremum is $\sigma^2 = \sum[Y_j - f_j(\hat{\boldsymbol{\theta}}_{(j)})]^2/n$. When testing models from $\mathcal{S}$ a DM based on this quasi-likelihood criterion selects the model associated with the greatest value of $\lambda^* = \prod_{j=1}^{n} d_j(\hat{\boldsymbol{\theta}}_{(j)})$.

In the nested model case where $f_i \subset f_q$, based on the DM of Geisser, we can test hypothesis (3.3.1) utilizing $\lambda^* = \lambda_i^*/\lambda_q^*$ where λ_r^* is the value of λ^* for model f_r, for $r = i$ or q. We reject H_0, thereby choosing f_q, if $\lambda^* < 1$ and accept H_0, or select f_i, when $\lambda^* \geq 1$. Stone (1977) has shown that under true model f_i, $\lambda^* \rightarrow \lambda \exp(m(q) - m(i))$ as $n \rightarrow \infty$ where $m(r)$ is the number of fitted parameters for $f_r, r = i, q$, and λ is the likelihood ratio of the GLRT given by (3.3.3). Thus, as $n \rightarrow \infty$ under regularity conditions the power of this test is $P[\chi^2 > 2(m(q) - m(i))]$ where χ^2 has degrees of freedom $m(q) - m(i)$ (see Problem 3.5).

In the framework of question A3, assuming a model from $\mathcal{S}$ is correct, consider the natural process of adding complexity to our models and thus obtaining $f_1 \subset f_2 \subset \cdots f_k$. We pairwise test hypotheses (3.3.1) using a test statistic, say T_i, for $i = 1, 2, \ldots, k-1$. Under normality of the error let $P(T_i > t_i \mid f_i \text{ correct}) = p_i$ for $1 \leq i \leq k-1$, where t_i is the observed value of T_i. An SM can be constructed by stopping and declaring f_i correct the first time $p_i > \alpha$ for some α, $0 < \alpha < 1$. This results in a "forward" selection procedure. Here, if f_i is selected then for $q < i$

$$P[\text{IS}(f_i \mid f_q)] \leq \alpha \tag{3.3.7}$$

We see that with this SM for α fixed the conditional probability of overfitting the model, that is, including more parameters than is necessary, is bounded above by α. However, with this forward selection procedure the probability of underfitting the model, that is, stopping too soon and thus excluding needed parameters, cannot be directly calculated. The topic of stepwise DMs for linear regression models is discussed in Section 3.8. In other cases selection probabilities have been explored, and we refer to Shibata (1976) and Woodroofe (1982) for specific examples.

The GLRT can be used to test a proposed model for lack of fit, or model misspecification. With this test, outlined in the following example, the set of proposed models $\mathcal{S}$ can be shrunk, in a manner similar to that with (2.1.1) and (3.1.2), to a manageable size.

Example 3.3.2

For a proposed model from $\mathcal{S}$, the model function can be expanded to form the new model function

$$g_j(\boldsymbol{\theta}, \boldsymbol{\beta}) = f_j(\boldsymbol{\theta}) + w_j(\boldsymbol{\beta}) \tag{3.3.8}$$

where $\boldsymbol{\beta}$ is a vector of unknown parameters. Clearly $f \subset g$ and (3.3.1) can be tested where f and g are assumed under H_0 and H_1, respectively. This test procedure is general and its nature depends on the formulation of $w_j(\boldsymbol{\beta})$. A particular construction of $w_j(\boldsymbol{\beta})$, proposed by Utts (1982), leads to a test of model function adequacy called the rainbow test. Here the data set $(\mathbf{Y}, \mathbf{x})$ is split into two parts, $(\mathbf{Y}^{(1)}, \mathbf{x}^{(1)})$ and $(\mathbf{Y}^{(2)}, \mathbf{x}^{(2)})$, where $w_j(\boldsymbol{\beta})$ is defined as zero for all $\mathbf{x}_j$ in $\mathbf{x}^{(1)}$ and β_j for $\mathbf{x}_j$ in $\mathbf{x}^{(2)}$. In this way RSS($\hat{g}$) is just the residual sum of squares of f restricted to $(\mathbf{Y}^{(1)}, \mathbf{x}^{(1)})$, which we denote RSS($\hat{f} \mid (\mathbf{Y}^{(1)}, \mathbf{x}^{(1)})$). The degrees of freedom associated with test statistic (3.3.3) are $n - n1$ where $(\mathbf{Y}^{(1)}, \mathbf{x}^{(1)})$ consists of $n1$ data points. As before, if the model function is nonlinear the GLRT based on (3.3.3) is approximate. When the model function is linear an exact test can be constructed. The specifics of this test are presented in Section 3.6.

3.4 MODEL STABILITY

Any DM used in the estimated model situation must assess the model's sensitivity to the observed set of responses. In this section, based on the asymptotic theory resulting from LS estimation, a DF which measures a model function's stability at each location is constructed. In the framework of Atkinson's question A2, choosing not to assume a true model exists in $\mathcal{S}$, this DF can be employed as statistical evidence which helps separate efficient from nonefficient models. this analysis occurs often in the social sciences, where the goal is a parsimonious model. Also, using this DF the design space, or region containing the settings of the predictor variables, can be partitioned into "stable" and "unstable" areas with respect to the individual model functions. In the first theorem the large sample variance of a fitted model function, ignoring bias (3.2.11), at any location is computed.

Theorem 3.4.1

For true model from $\mathcal{S}$ under NRC and (3.2.10) letting $\mathrm{Var}_b[\hat{f}_j] = E[\hat{f}_j - f_j]^2$ we find

$$\mathrm{Var}_b[\hat{f}_j] = \sigma^2\alpha_j + \sigma^4\Delta_j + o(n^{-1/2}) \tag{3.4.1}$$

as $n \to \infty$ where

$$\alpha_j = \mathbf{F}_j'\mathbf{M}\mathbf{F}_j \tag{3.4.2}$$

and

$$\Delta_j = \mathrm{Tr}[\mathbf{M}\mathbf{H}_j\mathbf{M}\mathbf{H}_j]/2 + \{\mathrm{Tr}[\mathbf{M}\mathbf{H}_j]\}^2/4 \tag{3.4.3}$$

for $1 \leq j \leq n$.

Proof We fix a model from $\mathcal{S}$ and employ (3.2.7). Now $[\hat{f}_j - f_j]^2$ is

$$\boldsymbol{\phi}'\mathrm{F}_j\mathrm{F}_j'\boldsymbol{\phi} + [\boldsymbol{\phi}'\mathrm{H}_j(\boldsymbol{\theta}^*)\boldsymbol{\phi}]^2/4 + \mathrm{F}_j'\boldsymbol{\phi}\boldsymbol{\phi}'\mathrm{H}_j(\boldsymbol{\theta}^*)\boldsymbol{\phi}/2 \tag{3.4.4}$$

Strong convergence along with the differentiability conditions implies convergence of $\boldsymbol{\theta}^*$ and $\mathrm{H}_j(\boldsymbol{\theta}^*)$. Using (3.2.10), the expected value of (3.4.4) becomes

$$E[\mathbf{V}'\mathbf{F}_j\mathbf{F}_j'\mathbf{V}] + E[\mathbf{V}'\mathbf{H}_j\mathbf{V}]^2/4 + E[\mathbf{F}_j'\mathbf{V}\mathbf{V}'\mathbf{H}_jV]/2 + o(n^{-1/2})$$

as $n \to \infty$, where $\mathbf{V} \sim N(0, \sigma^2\mathbf{M})$. The proof is completed using the expected value formulas for quadratic forms in normal variables given in Searle (1971, p. 66).

Using the asymptotic theory of LS estimators, an estimate of (3.4.1) is constructed. The "biased" variance of $\hat{f}_j$, approximated by (3.4.1), is estimated by

$$S_b^2(\hat{f}_j) = \hat{\sigma}^2\hat{\alpha}_j + \hat{\sigma}^4\hat{\Delta}_j \tag{3.4.5}$$

where

$$\hat{\alpha}_j = \hat{\mathbf{F}}_j'\hat{\mathbf{M}}\hat{\mathbf{F}}_j \tag{3.4.6}$$

and

$$\hat{\Delta}_j = \mathrm{Tr}[\hat{\mathbf{M}}\hat{\mathbf{H}}_j\hat{\mathbf{M}}\hat{\mathbf{H}}_j]/2 + \left\{\mathrm{Tr}[\hat{\mathbf{M}}\hat{\mathbf{H}}_j]\right\}^2/4 \tag{3.4.7}$$

In (3.4.5) $\hat{\sigma}^2$ is a consistent estimate, possibly obtained from repeat data points, of σ^2. We see that S_b^2 is the sum of two parts. The first term, $\hat{\sigma}^2\hat{\alpha}_j$, is a result of the linear part of the model function, while the second term, $\hat{\sigma}^4\Delta_j$, is needed because of the nonlinearity component.

We desire a DF which assesses a fitted model function's stability with an intuitive meaning. To do this, we first assume a correct model and examine the large sample distribution of $\hat{f}_j$. Combining

NRC and the bias calculation of Box yields the limiting distribution

$$L\left\{[\hat{f}_j - f_j]/\operatorname{Var}(f_j)^{1/2}\right\} \longrightarrow N(\beta_j, 1) \qquad (3.4.8)$$

as $n \to \infty$, where β_j is given by (3.2.11) and $\operatorname{Var}(f_j) = E[\hat{f}_j - f_j - \beta_j]^2$. Applying this distribution, an asymptotic upper confidence bound for the squared deviation of the fitted model function from the assumed true model function is derived. This is seen in the following theorem.

Theorem 3.4.2

Under (3.4.8) an asymptotic $1-\alpha$ upper confidence bound for $[\hat{f}_j - f_j]^2$ is $\operatorname{Var}_b(\hat{f}_j)\chi^2(1-\alpha, 1, \delta(j))$, that is,

$$\lim_{n\to\infty} P\left\{[\hat{f}_j - f_j]^2 \le \operatorname{Var}_b(f_j) \times \chi^2_{(1-\alpha,1,\delta(i))}\right\} \ge 1-\alpha \qquad (3.4.9)$$

where $\chi^2_{(1-\alpha,1,\delta(j))}$ is the $(1-\alpha)$100th percentile point of a noncentral chi-square random variable with one degree of freedom and noncentrality parameter $\delta(j) = \beta_j^2/\sigma^2$.

Proof Using (3.4.8) the large sample distribution of $[\hat{f}_j - f_j]^2/\operatorname{Var}(\hat{f}_j)$ is $\chi^2_{(1,\alpha(j))}$. Hence,

$$\lim_{n\to\infty} P\left\{[\hat{f}_j - f_j]^2 \le \operatorname{Var}(\hat{f}_j)\chi^2_{(1-\alpha,1,\delta(j))}\right\} = 1-\alpha$$

The proof is completed using the well-known property $E[f_j - f_j]^2 \ge E[\hat{f}_j - f_j - \beta_j]^2$ as $n \to \infty$.

For a proposed model and large n an approximate upper confidence limit for $[\hat{f}_j - f_j]^2$ is obtained by substituting the estimates for $\operatorname{Var}_b(\hat{f}_j)$ and β_j given by (3.4.5) and (3.2.12) into (3.4.9). This estimated bound is compared to a bound on the range of the responses, thereby producing an intuitive measure of a fitted model function's stability. Under normality of the responses the width of a confidence interval for $Y_j - E(Y_j)$, with confidence coefficient $1 - \alpha$, is $2\sigma z_{(1-\alpha/2)}$. Thus, for a fixed α the range on $|Y_j - E(Y_j)|$ is $\sigma z_{(1-\alpha/2)}$. A DF which assesses a model's stability at location $\mathbf{x}_j$, called the stability function or SF, is given by

$$\mathrm{SF}(\hat{f}_j) = [S_b^2(\hat{f}_j)\chi^{2'}_{(1-\alpha)}]^{1/2}/[\hat{\sigma} z_{(1-\alpha/2)}] \qquad (3.4.10)$$

where $S_b(\hat{f}_j)$ is defined by (3.4.5), $\hat{\sigma}^2$ is a consistent estimate of σ^2, and $\chi^{2'}$ is a noncentral chi-square random variable with one degree of freedom and noncentrality parameter $\delta(j) = \beta_j^2/\sigma^2$. In an asymptotic sense, SF is the ratio of the ranges of $\hat{f}_j$ to Y_j at x_j for $j \geq 1$.

In the case where the estimated bias of $\hat{f}_j$ is negligible, (3.4.10) can be simplified. If $\hat{\beta}_j = 0$, noting $[\chi^2_{(1-\alpha,1)}]^{1/2} = z_{(1-\alpha/2)}$, we find that (3.4.10) reduces to

$$\mathrm{SF}(\hat{f}_j) = [S_b(\hat{f}_j)/\hat{\sigma}^2]^{1/2} \qquad (3.4.11)$$

Further, when $\beta_j = 0$, then (3.4.11) will be smaller than (3.4.10). This implies that (3.4.11) can be used as a conservative measure of stability.

For comparison purposes we note that for a linear model with uncorrelated predictor variables $S_b^2 = \hat{\sigma}^2 m/n$ in the optimum setting (see Problem 3.7). Applying (3.4.11) in this case we find $\mathrm{SF}(\hat{f}_j) = (m/n)^{1/2}$. Hence, for a proposed model the gain in SF over and above $(m/n)^{1/2}$ indicates the magnitude of the instability at location $\mathbf{x}_j$.

Example 3.4.1

Consider three models with model functions given by $f_j(1,\theta_1) = x_j\theta_1, f_j(2,\theta_2) = \exp(x_j\theta_2) - 1$, and $f_j(3,\boldsymbol{\theta}) = x_j\theta_3\exp(x_j\theta_4)$ where $\boldsymbol{\theta}' = (\theta_3,\theta_4)$. Thus, $m(1) = m(2) = 1$ while $m(3) = 2$; model f_1 is a reduction of f_3 since $f_j(1,\theta_1) = f_j(3,\boldsymbol{\theta}^*)$ for $\boldsymbol{\theta}^{*\prime} = (\theta_1, 0)$. For this first model we find $F_j(1,\theta_1) = x_j$ and $H_j(1,\theta_1) = 0$ for $j \geq 1$. With the second model $F_j(2,\theta_2) = x_j\exp(x_j\theta_2)$ and $H_j(2,\theta_2) = x_j^2\exp(x_j,\theta_2)$. Based on the third model $F_j(3,\boldsymbol{\theta})' = (x_j\exp(x_j\theta_4), x_j^2\theta_3\exp(x_j\theta_4))$ and

$$H_j(3,\boldsymbol{\theta}) = \begin{pmatrix} 0 & x_j^2\exp(x_j\theta_4) \\ x_j^2\exp(x_j) & x_j^3\theta_3\exp(x_j\theta_4) \end{pmatrix}$$

For exposition, samples are drawn at $x_j = 0$, 1, and 2 and f_1 is taken to be correct with $\theta_1 = 1$. Here, model f_3 is admissible but, since $f_1 \subset f_3$, it is not considered the true one. The limiting model functions take parameter values which minimize the square deviations from the true model function. These values are $\theta_2 = .561$, $\theta_3 = 1$, and $\theta_4 = 0$. Listed in Table 3.2 are the limiting model function values.

Using (3.2.5), we find that the asymptotic variance for $\hat{\boldsymbol{\theta}}_1$ is $\sigma^2/\sum F_j(1)F_j(1)' = .2\sigma^2$ and for $\hat{\theta}_2$ is $.0245\sigma^2$. For $\hat{\boldsymbol{\sigma}}$, from the

Table 3.2 Limiting Model Values

	Locations		
Models	0	1	2
$f_j(1,1)$	0	1	2
$f_j(2,.561)$	0	.7524	2.0710
$f_j(3,(1,0)')$	0	1	2

third model, the asymptotic dispersion matrix is $\sigma^2\mathbf{M}(3)$ where

$$\mathbf{M}(3) = \begin{pmatrix} 4.25 & -2.25 \\ -2.25 & 1.25 \end{pmatrix}$$

It is instructive to note that multiplying the extraneous factor $\exp(x_j\theta_4)$ with $f_j(1,\theta_1)$, thereby constructing $f_j(3,\theta)$, increases the asymptotic variance of $\hat{\theta}_1$ 20 times, from $.2\sigma^2$ to $4.25\sigma^2$.

Based on the three model functions, $\hat{\beta}_j$ is calculated using (3.2.12) along with $\hat{\alpha}_j$ given by (3.4.6) and $\hat{\Delta}_j$ defined by (3.4.7) at each location with $\hat{\sigma}^2 = .5$. Further, we compute S_b^2 given by (3.4.5) and, noting the small biases found, calculate SF given by (3.4.11) for the models at each location. At $x = 0$ the fitted model functions are fixed, which results in zero calculations for these formulas. For $x = 1$ and 2 the pertinent computations are listed in Table 3.3.

We observe from Table 3.3 that models f_1 and f_2 are both fairly stable at each location since $(m/n)^{1/2} = .577$. In addition, $f_j(1)$ is linear, producing $\hat{\Delta}_j = 0$ for $j = 1, 2$, while $f_j(2)$ is slightly nonlinear with $\hat{\Delta}_2 = .001$ and $\hat{\Delta}_3 = .068$. In contrast, model $f_j(3)$ is highly unstable at both $x = 1$ and 2. In particular, when $x = 2$ the value

Table 3.3 Stability Assessment—Theoretical Models

Location	Model	$\hat{\beta}_j$	$\hat{\alpha}_j$	$\hat{\Delta}_j$	S_b^2	SF
$x = 1$	f_1	0	.2	0	.1	.447
	f_2	−.01	.075	.001	.038	.276
	f_3	0	1.0	8.17	2.54	2.255
$x = 2$	f_1	0	.8	0	.4	.894
	f_2	.003	.925	.068	.479	.979
	f_3	0	1.0	52.0	13.5	5.196

$SF(f_j(3)) = 5.196$ implies that the fitted model's effective range, through repeated sampling, is five times as large as that of the responses. Lastly, we note that at $x = 1$ an incorrect model, namely f_2, may be desired over the true one, f_1, since its asymptotic variance is smaller. This can be seen through the construction of large sample confidence regions (Problem 3.8).

For a proposed model from $\mathcal{S}$ the entire design space may be restricted, in terms of location variables $\mathbf{x}_j$, to a region where the model function values are stable. In fact, if the set of competing model functions fit the responses equally well (see Section 3.5) a reasonable DM is to select the model with the largest region of stable locations. This section is concluded with an example dealing with agricultural growth models.

Example 3.4.2

This example, taken from Ratkowsky (1983, pp. 61-75), concerns sigmoidal or S-shaped growth curves. Three models relating a response variable to a set of predictor variables are projected for a data set. First the Gompertz model, denoted f_1, has model function

$$f_j(1, \boldsymbol{\theta}) = \theta_1 \exp[-\exp(\theta_2 - \theta_3 x_j)]$$

where $\boldsymbol{\theta}' = (\theta_1, \theta_2, \theta_3)$. Second, the logistic model, f_2, is given by model function

$$f_j(2, \boldsymbol{\theta}) = \theta_1 / [1 + \exp(\theta_2 - \theta_3 x_j)]$$

where $\boldsymbol{\theta} = (\theta_1, \theta_2, \theta_3)$. Last, the Richards (1959) model, f_3, has model function

$$f_j(3, \boldsymbol{\theta}) = \theta_1 / [1 + \exp(\theta_2 - \theta_3 x_j)]^{1/\theta_4}$$

Table 3.4 Sigmoidal Data—Y: Growth, x: Time

x	y	x	y
9	8.93	57	56.11
14	10.80	63	61.73
21	18.59	70	64.62
28	22.33	79	67.08
42	39.35		

for $\boldsymbol{\theta}' = (\theta_1, \theta_2, \theta_3, \theta_4)$. For the first two models $m = 3$, while the last model has $m = 4$. These models are applied to data (unpublished) from W. F. Hunt for pasture growth Y against time x. The data are listed in Table 3.4. The LS estimates are calculated along with the RSS and these appear in Table 3.5. For all the proposed models R^2, defined by (3.1.1), and the DF measuring total significance, namely (3.3.4), yield little discriminating information. Here, SGF(f) > .991 and $R^2(f) > .995$ for all the models. In addition, listed in Table 3.5 is $S^2(f)$ given by (3.3.5) for these models.

Based on Table 3.5, we find that the model with the smallest RSS and $S^2(f)$ values is, predictably, the one with the greatest number of parameters, namely f_3. To measure the effects of the number of estimated parameters, along with the forms of the model

Table 3.5 Sigmoidal Model Statistics

Model	Estimates				Statistics	
	$\hat{\theta}_1$	$\hat{\theta}_2$	$\hat{\theta}_3$	$\hat{\theta}_4$	RSS	S_f^2
f_1	82.83	1.22	.037	—	21.770	3.63
f_2	72.46	2.62	.067	—	8.019	1.34
f_3	69.62	4.26	.089	1.724	6.010	1.20

functions, SF given by (3.4.11) is computed at all locations using $\hat{\sigma}^2 = 1.3$. We remark that for all models at all locations the amount of bias was minimal with $|\hat{\beta}_j| \leq .19$. The SF calculations are presented in Table 3.6.

On viewing Table 3.6, the Gompertz and logistic models appear more stable than the Richards model. For f_1 and f_2 the fitted model function values, as measured by SF, deviate about half as much as the observed growth. In contrast, for f_3 the estimated model function's variability approaches, for all but the lower three values of x, the observed growth. Thus, based on this data set the Richards model might be eliminated from consideration in favor of either of the other two S-shaped growth models. In the event the Richards model is desired, based on theoretical considerations, the poor stability indicates that either more samples are required or the model should be restricted to smaller values of x ($x \leq 21$).

We would be remiss if we did not note that the analysis in Example 3.4.2 is dependent on the hypothesized value of σ^2. A well-constructed experiment should contain repeat data points, which enables calculation of an estimate of σ^2 independent of any proposed model function (note (2.2.6)). However, an initial estimate of the variability of the responses may be utilized in the formulas measuring stability, namely SF, for comparison of the competing models. This is precisely what is done in the previous example.

Table 3.6 Stability Assessment—SF Values

Model	Locations								
f_1	.56	.56	.52	.52	.67	.61	.58	.63	.86
f_2	.49	.52	.52	.53	.68	.60	.52	.57	.79
f_3	.85	.84	.89	1.10	1.65	1.56	1.38	1.19	1.19

3.5 MODEL ACCURACY

In this section we consider the characteristic of an estimated model function that, in conjunction with an assessment of stability, signals its usefulness in modeling an observed process. As in the previous section, the asymptotic theory of LS estimation is used to construct a DF which assesses a fitted model function's accuracy or closeness to responses at individual locations. Based on question A2, this measurement offers evidence in the model discrimination process. Similar to the DF for model stability, SF, such an assessment statistic can be used to designate "outlier" responses and locate regions of the design space where a proposed model function may not be applicable. In this section we extend the notion of studentized residuals to the nonlinear model function setting. Overall measures of a proposed model's accuracy are left to the next section.

Any construction of a DF to detect outliers, or responses that have means at values other than those taken by an assumed fitted model function, must start with a measure of the variance of the model's residuals. In the following theorem we give, for a proposed model, the asymptotic variance of the residuals at individual locations.

Theorem 3.5.1

Fix a model from $\mathcal{S}$ given by (3.2.1). Under NRC and (3.2.10) we have the approximations

$$E[Y_j - \hat{f}_j] = -\beta_j \tag{3.5.1}$$

where β_j is given by (3.2.11) and for large n

$$E[Y_j - \hat{f}_j]^2 = \sigma^2(1-\alpha_j) - 3\sigma^4\Delta_j + o(n^{-1/2}) \tag{3.5.2}$$

where α_j and Δ_j are defined by (3.4.2) and (3.4.3), respectively, for $1 \leq j \leq n$.

Proof Fix a model from $\mathcal{S}$. First, (3.5.1) is a direct consequence of the convergence of the LS estimators. Second, to obtain (3.5.2) we expand $[Y_j - f_j]^2$ by adding and subtracting $\hat{f}_j$ to obtain

$$[Y_j - \hat{f}_j]^2 = [Y_j - f_j]^2 - [\hat{f}_j - f_j]^2 - 2[Y_j - \hat{f}_j][\hat{f}_j - f_j] \qquad (3.5.3)$$

where f_j is the true model function value at $\mathbf{x}_j$, that is, $E(Y_j) = f_j = f_j(\boldsymbol{\theta})$ for $\boldsymbol{\theta} \in R^m$. The expected value of (3.5.3) is now taken. Under the true model the expected values of the first and second terms are σ^2 and (3.4.1), respectively. Utilizing (3.2.7), the third term's value is twice

$$\mathbf{e}_j\mathbf{F}_j'\boldsymbol{\phi} + \mathbf{e}_j\boldsymbol{\phi}'\mathbf{H}_j(\boldsymbol{\theta}^*)\boldsymbol{\phi}/2 - \boldsymbol{\phi}'\mathbf{F}_j\mathbf{F}_j'\boldsymbol{\phi} - [\boldsymbol{\phi}'\mathbf{H}_j(\boldsymbol{\theta}^*)\boldsymbol{\phi}]^2/4 - 2\mathbf{F}_j'\boldsymbol{\phi}\boldsymbol{\phi}'\mathbf{H}_j(\boldsymbol{\theta}^*)\boldsymbol{\phi}/2 \qquad (3.5.4)$$

Employing (3.2.10), we find

$$E[\mathbf{e}_j\mathbf{F}_j'\boldsymbol{\phi}] = \sigma^2\mathbf{F}_j'\mathbf{M}\mathbf{F}_j + o(n^{-1/2}) \qquad (3.5.5)$$

and

$$E[\boldsymbol{\phi}'\mathbf{F}_j\mathbf{F}_j'\boldsymbol{\phi}] = \sigma^2\,\mathrm{Tr}[\mathbf{M}\mathbf{F}_j\mathbf{F}_j'] + o(n^{-1/2}) \qquad (3.5.6)$$

The difference between (3.5.5) and (3.5.6) is of the order $o(n^{-1/2})$, and hence their combined effect in (3.5.4) is negligible. Finally, under NRC, taking the expected value of (3.5.4) and dropping $E[\mathbf{F}_j'\boldsymbol{\phi}\boldsymbol{\phi}'\mathbf{H}_j(\boldsymbol{\theta}^*)\boldsymbol{\phi}/2]$ and $E[\mathbf{e}_j\boldsymbol{\phi}'\mathbf{H}_j(\boldsymbol{\theta}^*)\boldsymbol{\phi}/2]$, since they have asymptotic value $o(n^{-1/2})$, yields the desired result.

Higher-order, and hence more accurate, approximations for (3.5.2) exist, but the one presented is proposed because of its accuracy and ease of calculation. With the help of the previous theorem a DF used to evaluate an estimated model function's fit, or accuracy, at any location in the design space is constructed. For a proposed model substituting LS estimators we form the accuracy DF given by

$$\mathrm{AF}(\hat{f}_j) = [Y_j - \hat{f}_j - \hat{\beta}_j]/\mathrm{se}[Y_j - \hat{f}_j] \tag{3.5.7}$$

where $\hat{\beta}_j$ is defined by (3.2.12) and

$$\mathrm{se}[Y_j - \hat{f}_j] = [\hat{\sigma}^2(1 - \hat{\alpha}_j) - 3\hat{\sigma}^4\hat{\Delta}_j]^{1/2}$$

As before, $\hat{\sigma}^2$ is a consistent estimate of σ^2 and $\hat{\alpha}_j$ and $\hat{\Delta}_j$ are defined by (3.4.6) and (3.4.7), respectively.

Under consistency conditions the limiting distribution of (3.5.7) is standard normal, and hence outliers are determined by comparisons with standard normal percentile values. Based on (3.5.7) a DF measuring fit at $\mathbf{x}_j$ is

$$\mathrm{PAF}(\hat{f}_j) = 2P[Z > |\mathrm{AF}(\hat{f}_j)|] \tag{3.5.8}$$

For small values of PAF a model's fit is suspect. Also, because of the asymptotic nature of the derivation, AF may not be defined, that is, $\hat{\sigma}^2(1 - \hat{\alpha}_j) < 3\hat{\sigma}^4\hat{\Delta}_j$. In this case more sample points are needed to ensure proper convergence; however, the modeler can utilize simulation measures to test for lack of fit (see Chapter 4). Following a remark, the proposed accuracy assessment statistic is applied to the sigmoidal growth models introduced in the previous section.

Remark 3.5.1

In the linear model case the DF defined by (3.5.7) is referred to as a studentized residual (see Myers, 1986, p. 144). In this setting for a model from $\mathcal{S}$, α_j is the jth diagonal element in the "hat" matrix $\mathbf{x}[\mathbf{x}'\mathbf{x}]^{-1}\mathbf{x}'$ and represents the variance of the fitted model function value $\hat{f}_j$ for $j \geq 1$. Using (3.5.2), we can extend this idea of quantifying a particular data point's effect on the estimated model function to nonlinear model functions. The leverage of $\mathbf{x}_j$ on the estimated value $\hat{f}_j$ is

$$\mathrm{LV}(\hat{f}_j) = \hat{\alpha}_j + 3\hat{\sigma}^2\hat{\Delta}_j \tag{3.5.9}$$

where $\hat{\sigma}^2$ is an independent estimate of σ^2. Large LV values indicate a large influence by $\mathbf{x}_j$.

Example 3.5.1

The models and data introduced in Example 3.4.2 are examined in the context of model function fit. Applying (3.5.7), under the three models, we search for locations where the estimated model function's closeness to the response variable is in question. The three models are examined for accuracy of fit. The high instability of the Richards model, f_3, results in AF being undefined at all the locations. In Table 3.7 are listed the values of AF for the competing models f_1 and f_2 at each location using $\hat{\sigma}^2 = 1.3$. In comparing the model's AF values we find that f_2 has fewer atypical or possible outlier responses than f_1. In view of the large sample normality

Table 3.7 Accuracy Assessment—AF Values

Model	Locations								
f_1	1.72	−.19	1.24	−2.54	−1.42	1.11	2.30	−.06	DNE
f_2	.37	−.66	1.92	−1.20	− .79	.02	1.11	−.09	−1.57

of AF, we note that for the Gompertz model, f_1, three out of the nine locations have small PAF values (one is undefined because of instability and the other two are less than .025). This indicates problems of fit at those locations. The logistic model, f_2, is superior in terms of its fit to the growth data, with only one location, namely $\mathrm{AF}(\hat{f}_j(2)) = 1.92$ and $\mathrm{PAF}(\hat{f}_j(2)) = .05$ for $j = 3$, suspect. This may signal the beginning of change in the S-shaped growth curve or may simply be due to random error.

Remark 3.5.2

Rigorous procedures for detecting outliers in proposed models have been proposed in the literature and for an overview we refer to John and Draper (1978), Rosner (1983), and Kitagawa (1979). These methods include the designation of the number of outlier points to be sought after as well as a lack-of-fit measurement.

As mentioned earlier, any model comparison involving fitted model functions should be viewed in the context of stability and accuracy assessment. The proposed statistics for measuring these basic characteristics, presented in this and the previous section, compliment another and are to be used to gain an understanding of the rival models, in particular their problems. In the framework of question A3, many DMs which utilize SF and AF, in various degrees, to eliminate unlikely models or select the best representative from S can be constructed. Here, even though selection probabilities are not calculated or used, model choices can still be made.

The model assessment statistics SF and AF are asymptotic in nature and for small or even moderately sized samples may not be applicable. For small sample sizes AF may not even be defined. For any particular problem the characteristic of these DFs can be verified by resampling procedures. This topic is explored in the next chapter.

3.6 CRITERION FUNCTIONS

In the case of estimated model functions we have seen that there exist two basic properties, namely stability and accuracy. The previous two sections of this chapter were devoted to the development of DFs which assess these traits. In this section we consider the construction of a single DF, called a criterion function or CF, which takes into account the prospective model function's fit and stability. Here, under Atkinson's question A2, DMs that are based on the CF values for the rival models are examined. These CFs are used to eliminate unlikely models or select a best model from S without evaluation by IS probabilities.

We start our investigation with a class of DFs which are a form of the penalized likelihood function as described by Good and Gaskins (1971). For $f_i \in S$ the general form of CF is given by

$$\mathrm{CF}_g(\hat{f}_i) = \mathrm{MRSS}(\hat{f}_i) + \lambda J(\hat{\boldsymbol{\theta}}) \tag{3.6.1}$$

for constant λ, where $J(\boldsymbol{\theta})$ is a function of parameter $\boldsymbol{\theta}$ or an appropriate estimator. We see that CF_g is the sum of two terms. The first term, MRSS, measures the accuracy of the fitted model function to the responses, while the second term, $\lambda J(\hat{\boldsymbol{\theta}})$, is a penalty function which penalizes the estimated model according to its instability. The larger the model's instability, the larger $J(\hat{\boldsymbol{\theta}})$ should be. Since $J(\hat{\boldsymbol{\theta}})$ is frequently a function of the variance of the responses, $J(\hat{\boldsymbol{\theta}})$ often contains $\hat{\sigma}^2$, a consistent estimator of σ^2. Hence, CF_g balances the two basic model characteristics as dictated by λ. Also, for penalty function $J(\hat{\boldsymbol{\theta}})$ and

λ the DM which selects the model, $f_i \in \mathcal{S}$, associated with the least CF_g value is called the least criterion method, denoted LCM.

We remark that in some instances CF_g, given by (3.6.1), is an extension of QDF, defined by (1.4.3). This occurs when the penalty function $J(\boldsymbol{\theta})$ is quadratic in $\boldsymbol{\theta}$. For other cases we refer to Problem 3.10.

Since the penalty function is to be sensitive to the stability of the proposed models, we consider DFs of the form (3.6.1) where $J(\hat{\boldsymbol{\theta}})$ is a function of the biased variance, S_b^2. From (3.4.5) we note that S_b^2 can be partitioned into linear and nonlinear parts and, hence, $J(\hat{\boldsymbol{\theta}})$ is taken to be a function of $\hat{\alpha}_j^2$ and $\hat{\Delta}_j^4$ for $1 \leq j \leq 1$. Since the penalty function is cumulative with respect to the locations we find that, for a proposed model, $J(\hat{\boldsymbol{\theta}})$ is to be a function of $\sum \alpha_j = \mathrm{Tr}[\hat{\mathbf{M}}\hat{\mathbf{F}}'\hat{\mathbf{F}}] = m$, the number of estimated parameters, and $\sum \hat{\Delta}_j$. Thus, the linear part of the penalty is a function of the number of unknown parameters. Keeping in mind that $J(\hat{\boldsymbol{\theta}})$ should increase as the number of parameters increases, the linear penalty portion can be written as $-\hat{\sigma}^2[n-m]/n$. For $f_i \in \mathcal{S}$ we consider CFs of the form

$$\mathrm{CF}(\hat{f}_i) = \mathrm{RSS}(\hat{f}_i) - \hat{\sigma}^2[n-m] + \lambda\hat{\sigma}^4\hat{\Delta} \qquad (3.6.2)$$

where λ is a constant and $\hat{\Delta} = \sum \hat{\Delta}_j$. With this construction, under true linear model $f_i(\hat{\Delta} = 0)$, $E[\mathrm{CF}(\hat{f}_i)] = 0$ (see Problem 3.9). From (3.6.2) we note that CF is a combination of three parts: accuracy as measured by RSS, linear stability assessed by $\hat{\sigma}^2[n-m]$, and nonlinear stability accounted for by $\hat{\sigma}^4\hat{\Delta}$.

For a competing set of linear models, CF given by 3.6.2) can be simplified. In this case CF for $f_i \in \mathcal{S}$ becomes

$$\mathrm{CF}_L(\hat{f}_i) = \mathrm{RSS}(\hat{f}_i) - \hat{\sigma}^2[n - m] \qquad (3.6.3)$$

Under the true model CF_L should be close, as dictated by random error, to zero. Thus, under Atkinson's A2, CF_L can be used to designate unlikely models (those with large values of (3.6.3)). This procedure will be demonstrated in Example 3.6.2, when CF_L is extended to the situation where a consistent estimate of the variance does not exist.

Remark 3.6.1

When an independent (of any proposed model function) estimate of the variance σ^2 exists, the celebrated C_p statistic of Mallows (1973) for linear models takes the form

$$C_p(\hat{f}_i) = \mathrm{RSS}(\hat{f}_i)/\hat{\sigma}^2 - [n - 2m] \qquad (3.6.4)$$

for f_i. We note that C_p is a translation and therefore is equivalent for discrimination purposes to CF_L, $\hat{\sigma}^2\,\mathrm{CF}_L(\hat{f}_i) = C_p(\hat{f}_i) - m$, and it follows that for a true linear model function C_p should be about the number of estimated parameters.

In the general nonlinear model case the nonlinearity of the proposed model functions must be assessed in CF. How are we to define λ in (3.6.2)? It seems natural to extend the ideas of CF_L to this setting and find the value of λ which makes CF, under the true model, an asymptotically unbiased estimator of zero. In the following two lemmas and resulting theorem this value is obtained. Central in this exploration are the random variables

$$W_1(\boldsymbol{\theta}) = (\boldsymbol{\phi}'\mathrm{H}_j(\boldsymbol{\theta})\boldsymbol{\phi})_n^2 \qquad (3.6.5)$$

and

$$W_2(\boldsymbol{\theta}) = (\boldsymbol{\phi}'\mathbf{F}_j, \boldsymbol{\phi}'\mathbf{H}_j(\boldsymbol{\theta})\boldsymbol{\phi})_n \tag{3.6.6}$$

based on a prospective model from $\mathcal{S}$.

Lemma 3.6.1

Fix a model from $\mathcal{S}$ and let $b_{nrs,pq}(\boldsymbol{\theta}) = (f_{jrs}(\boldsymbol{\theta}),f_{jpq}(\boldsymbol{\theta}))_n$ and $b_{nr,sp}(\boldsymbol{\theta}) = (f_{jr}(\boldsymbol{\theta}),f_{jsp}(\boldsymbol{\theta}))_n$. Using the previous notation, evaluation at the true parameter $\boldsymbol{\theta}_0$, yields $b_{nrs,pq}$ and $b_{nr,sp}$. Assuming NRC, let $b_{nrs,pq} \to b_{rs,pq}$ and $b_{nr,sp} \to b_{r,sp}$, implying the convergence of the estimators $\hat{b}_{nrs,pq} \to b_{rs,pq}$ and $\hat{b}_{nr,sp} \to b_{r,sp}$ with probability one as $n \to \infty$. Let $\mathbf{V}' = (V_1, \ldots, V_m) \sim N(\mathbf{0}, \sigma^2\mathbf{A}^{-1})$. If $\boldsymbol{\theta}^* \to \boldsymbol{\theta}_0$ with unit probability as $n \to \infty$ then

$$L[n^1 W_1(\boldsymbol{\theta}^*)] \longrightarrow \sum\sum\sum\sum V_r V_s V_p V_q b_{rs,pq} \tag{3.6.7}$$

and

$$L[n^{1/2} W_2(\boldsymbol{\theta}^*)] \longrightarrow \sum\sum\sum V_r V_s V_p b_{rs,sp} \tag{3.6.8}$$

as n.

Proof Substituting $\boldsymbol{\theta}^*$ for $\boldsymbol{\theta}$ in (3.6.5) yields

$$W_1(\boldsymbol{\theta}^*) = \sum\sum\sum\sum \phi_r\phi_s\phi_p\phi_q(f_{jrs}(\boldsymbol{\theta}),f_{jpq}(\boldsymbol{\theta}^*))_n$$

Employing (3.2.5) and Lemma 3.2.1 yields (3.6.7). In a similar manner (3.6.8) is obtained.

Lemma 3.6.2

Under true model, NRC, and uniform integrability of the LS estimator,

$$E[n^1 W_1/4] = \sum_{j=1}^{n} \hat{\Delta}_j + o(1) \tag{3.6.9}$$

and

$$E[n^{1/2} W_2] = o(1) \tag{3.6.10}$$

where unit probability as $n \to \infty$, where $\hat{\Delta}_j$ is defined by (3.4.3) and the random variables W_r are evaluated for $r = 1, 2$.

Proof The proof is a result of (3.6.7), (3.6.8), and, to ensure convergence of moments, the uniform integrability condition.

We remark that relations (3.2.10) implies the uniform integrability requirement of Lemma 3.6.2. The proposed CF, in terms of the required λ value, is given in the following theorem.

Theorem 3.6.1

Fix $f_i \in \mathcal{S}$ and let the conditions of Lemmas 3.6.1 and 3.6.2 hold. If f_i is correct, then using $\lambda = 3, n^{-1}E[\text{CF}(f_i)] \to 0$, as $n \to \infty$.

Proof Under true model f_i from (3.2.13)

$$E[\text{MRSS}(\hat{f}_i)] = \sigma^2 - E[\text{Var}(\hat{f}_i)_n] + 2E[\text{Cov}(Y - \hat{f}_i, \hat{f}_i)_n] \tag{3.6.11}$$

as $n \to \infty$. Summing (3.4.4) over all design locations yields

$$\text{Var}(\hat{f}_i)_n = \boldsymbol{\phi}'\text{F}(\boldsymbol{\theta})\text{F}(\boldsymbol{\theta})\boldsymbol{\phi}/n + W_1(i, \boldsymbol{\theta}^*)/4 + W_2(\boldsymbol{\theta}^*)/2$$

where $\theta^* \to \theta_0$ with probability one as $n \to \infty$. From (3.6.9) and (3.6.10) it follows that

$$E[\mathrm{Var}(\hat{f}_i)_n] = m\sigma^2/n + \sigma^4\Delta/n + o(n^{-1}) \tag{3.6.12}$$

as $n \to \infty$. Similarly,

$$\mathrm{Cov}(Y - \hat{f}_i, \hat{f}_i)_n = -\boldsymbol{\phi}'\mathbf{F}'\mathbf{F}\boldsymbol{\phi}/n - W_1(\boldsymbol{\theta}^*)/4 - W_2(\boldsymbol{\theta}^*) + \mathbf{e}_j\boldsymbol{\phi}'\mathbf{H}_j(\boldsymbol{\theta}^*)\boldsymbol{\phi}/(2n) + e_j\mathbf{F}_j'\boldsymbol{\phi}/n + \Sigma e_j f_j/n$$

As a result of NRC and (3.2.10) and noting the asymptotic equivalence of (3.5.5) and (3.5.6), we have

$$E[\mathrm{Cov}(Y - \hat{f}_i, \hat{f}_i)_n] = \hat{\sigma}^4\hat{\Delta}/n + o((n^{-1/2}) \tag{3.6.13}$$

The asymptotic expected value of (3.6.3) with $\lambda = 3$, namely zero, is obtained by the substitution of (3.6.12) and (3.6.13) into (3.6.11).

Based on the previous theorem the CF, defined by (3.6.2) using $\lambda = 3$, can be used in an LCM. If either an incorrect model is examined, causing RSS to be large in relation to its number of fitted parameters, or the magnitude of stability in terms of its nonlinear portion is unsuitable, then CF will tend to be positive. In conjunction with such an LCM selection probabilities are hard to calculate and are left to simulation evaluation. Thus, in the framework of A2, models with large CF values relative to their competitors are eliminated from consideration.

Example 3.6.1

The LCM is applied to the models and data of Example 3.4.2. Using $\hat{\sigma}^2 = 1.3$, $\hat{\Delta}$ and CF given by (3.6.2.) with $\lambda = 3$ are calculated for the three models. Listed in Table 3.8 along with these

Table 3.8 Nonlinear Model Discrimination—CF

	f_1	f_2	f_3
m	3	3	4
$\hat{\Delta}$	.373	.079	7.202
CF	15.865	.951	36.023

values is m, the number of estimated parameters for the models. Here Δ accounts for the amount of nonlinear instability in a fitted model. From Table 3.8 we observe that for the model with the largest nonlinear instability, f_3, $\hat{\Delta}$ is relatively large. Because of this, f_3, although possessing the closest fit to the responses (note the minimum RSS and $S^2(f)$ values from Table 3.4), is judged by CF to be weaker than the others. The most desirable, in terms of CF, is model f^2, which possesses good fit and stability. This finding is consistent with the results of Example 3.4.2 for stability and Example 3.5.1 for accuracy. It is interesting to note that CF is not greatly affected by a few "bad" locations.

For nonlinear model functions, LCM, utilizing (3.6.2) with $\lambda = 3$, has the desirable property of searching for models with good fit in relation to their complexity. The complexity of any fitted model function is a combination of the number of estimated parameters and nonlinear stability as measured by $\hat{\Delta}$. In the next section we explore the relation of $\hat{\Delta}$ to the amount of nonlinearity in the proposed model f_i.

We now consider a CF based on the adjusted R^2 DF defined in (3.1.1). In the linear model function situation an adjusted R^2, accounting for the number of estimated parameters, has been proposed (see Draper and Smith, 1981, p. 92). For linear model $f_i \in S$

this DF is

$$R_a^2(\hat{f}_i) = 1 - (n-1)\,\mathrm{RSS}(\hat{f}_i)/[(n-m)\,\mathrm{RSS}(\bar{Y})] \qquad (3.6.14)$$

where m is the number of estimated parameters including the intercept. When R_a^2 is used as a CF, Kennard (1971) states that the resulting DM is closely related to the DM based on the minimum C_p criterion. In the discussion that follows the asymptotic forms introduced in Section 3.2 are employed to extend (3.6.14) to the nonlinear setting.

Let $f_i \in \mathcal{S}$ be the true model. Using (3.6.11) and Theorem 3.6.1, the adjusted R^2 DF in the nonlinear case is

$$R_a^2(f_i) = 1 - (n-1)[\mathrm{RSS}(\hat{f}_i) + 3\hat{\sigma}^4\hat{\Delta}]/[(n-m)\,\mathrm{RSS}(\bar{Y})] \quad (3.6.15)$$

where $\hat{\Delta} = \sum \hat{\Delta}_j$. It is clear that (3.6.15) reduces to (3.6.14) when f_i is linear ($\hat{\Delta} = 0$). Using (3.5.2) and (3.2.13) (substituting Y for f_2, noting $\mathrm{Var}(\bar{Y})_n = \sigma^2/n$, and assuming $\mathrm{Cov}(Y - Y, Y)_n \to 0$ as $n \to \infty$) we find

$$\lim R_a^2(\hat{f}_i) \longrightarrow \alpha/(\alpha + 1 + 1/n) \qquad (3.6.16)$$

where $\alpha = \delta_n/\sigma^2$. In this limiting form we find that R_a^2 is independent of the number of fitted parameters.

Example 3.6.2

In this example we analyze the Hald data listed in Draper and Smith (1981, pp. 629-66). There are four predictor variables and one response variable measured at 13 locations. The set of all possible linear regression models (see Example 1.3.1) constitutes $\mathcal{S}$. Since the true model is assumed to be contained in $\mathcal{S}$, an unbiased estimator of σ^2 is based on the full model f_p. The estimator is

$S^2(\hat{f}_p) = \text{RSS}(\hat{f}_p)/(13-4)$, which is observed to be 5.9829. Substitution of this estimator into CF_L, given by (3.6.3), yields for $f_i \in S$ the CF

$$\text{CF}_L(\hat{f}_i) = [S^2(\hat{f}_i) - S^2(\hat{f}_p)](n - m(i))/n \tag{3.6.17}$$

where $m(i)$ is the number of independent variables of f_i plus one. The values of CF_L along with R^2 and R_a^2, given by (3.1.1) and (3.6.14), are found for each of the competing models and are listed in Table 3.9 under the predictor variables utilized.

We observe that CF_L is more discriminating than the others because both R^2 and R_a^2 tend to increase with the number of fitted parameters. The models consisting of predictor variable combinations x_1x_2, $x_1x_2x_3$, $x_1x_2x_4$, and $x_1x_3x_4$ have CF_L values close to zero relative to their competitors. Using the minimum CF_L criterion, the linear model utilizing x_1x_2 is chosen. We remark that this is not the case if either R^2 or R_a^2 is employed as a CF.

The sampling properties of the difference in competing models, CF values are not easily deduced. For this reason DMs which se-

Table 3.9 Multiple Regression Models

Predictor variables	CF_L	R^2	R_a^2	Predictor variables	CF_L	R^2	R_a^2
1	92.3	.53	.49	2,3	27.3	.85	.82
2	64.6	.66	.64	2,4	62.2	.68	.62
3	144.	.13	.22	3,4	8.92	.94	.92
4	62.9	.67	.65	1,2,3	−.44	.98	.98
1,2	−.15	.98	.97	1,2,4	−.45	.98	.98
1,3	89.8	.54	.46	1,3,4	−.23	.98	.98
1,4	1.15	.97	.97	2,3,4	1.54	.97	.96
				1,2,3,4	0.0	.98	.98

lect as the true model the one with the smallest value of CF, CF_L, C_p, or R_a^2 have not been widely expounded. These CFs have been proposed for use in eliminating unlikely models in the spirit of Atkinson's question A2.

This section is concluded with a discussion of a test of model adequacy introduced in Example 3.3.2. The rainbow test of model function adequacy without repetition of locations in the case of linear models is examined. For a prospective model from $\mathcal{S}$, under the previous notation, the test function is

$$F(f_i) = (n1 - m)[\mathrm{RSS}(\hat{f}_i) - \mathrm{RSS}(\hat{f}_i \mid (\mathbf{Y}^{(1)}, \mathbf{x}^{(1)}))]/[(n - n1)\,\mathrm{RSS}(\hat{f}_i \mid (\mathbf{Y}^{(1)}, \mathbf{x}^{(1)}))] \qquad (3.6.18)$$

where under H_0: True Model f_i, F is distributed as an F-random variable with degrees of freedom $n - n1$ and $n1 - m$ (see Problem 3.25). Clearly, the form of the partition $(Y^{(1)}, x^{(1)})$ affects the distribution of F, and hence its power. Utts (1982) suggests that about half of the sample points be allocated to each group with points corresponding to high "leverage" put into $(\mathbf{Y}^{(2)}, \mathbf{x}^{(2)})$. In the nonlinear case this corresponds to choosing the points with the largest LV values, as defined by (3.5.7), to be eliminated under the alternative model (for $(\mathbf{Y}^{(2)}, \mathbf{x}^{(2)})$. Examples of this test are given in Utts (1982) and a specific problem is outlined in Problem 3.26.

Tests of model function adequacy, such as those given by LOF ((2.1.1) or (3.1.2)) and the rainbow test, can be used as CFs. However, they are best suited for preliminary testing of the models of $\mathcal{S}$ or justifying a model selected formally by a DM.

3.7 LINEARIZATION OF MODELS

In the previous section we saw that the nonlinear stability portion of CF is measured by $\hat{\Delta}$. In this section we explore the conditions under which an estimated model function may be efficiently approximated by a linear, in parameters, regression function. The appropriateness of such an approximation, referred to as linearizability, is seen to be a function of $\hat{\Delta}$. Hence, the ability of a model function to be linearized is tied to its stability. As we have demonstrated, for linear model functions the sum of the variances of the fitted model function values is σ^2 times the number of estimated parameters.

Under favorable conditions an estimated nonlinear model function may be expanded in the form of a Taylor series and the resulting model function used as an approximation in the range of the design space. Under a true model from $\mathcal{S}$ with correct parameter $\boldsymbol{\theta}_0$, using (3.2.7),

$$\hat{f}_j = \gamma_j + \mathbf{F}_j'\boldsymbol{\phi} + \boldsymbol{\phi}'\mathbf{H}_j(\boldsymbol{\theta}^*)\boldsymbol{\phi}/2 \tag{3.7.1}$$

where $\gamma_j = f_j - \mathbf{F}_j'\boldsymbol{\theta}_0$ and $\boldsymbol{\theta}^*$ is a suitable vector in R^m. At location $\mathbf{x}_j$ a linear approximation takes the form

$$f_j^L = c_j + \mathbf{a}_j'\hat{\boldsymbol{\theta}} \tag{3.7.2}$$

for constant c_j and $m \times 1$ vector of constants $\mathbf{a}_j$ for $j \geq 1$. We note that for large sample sizes if $\boldsymbol{\phi}'\mathbf{H}_j(\boldsymbol{\theta})\boldsymbol{\phi}$ acts as the constant zero in a neighborhood of $\boldsymbol{\theta}_0$, then there exists a linear approximation for f_j. Under this condition, assuming a large enough sample to ensure convergence of $\hat{F}_j$, taking $\mathbf{a}_j = \hat{\mathbf{F}}_j$ in (3.7.1) yields (3.7.2). Using resampling procedures, convergence properties of this type can be verified or tested (see Chapter 4). Hence, when proper conditions are satisfied, the linear approximation is obtained by the regression of Y_j on $\mathbf{a}_j = \hat{\mathbf{F}}_j$ including an intercept term c_j for $1 \leq j \leq n$. A theorem dealing with the linearizability of a proposed model function now follows.

Theorem 3.7.1

Fix a true model from $\mathcal{S}$ and assume NRC and (3.2.10). As $n \to \infty$ if $\sigma^2 \Delta_j \leq c$, for constant $c > 0$, then

$$E[\boldsymbol{\phi}'\mathbf{H}_j\boldsymbol{\phi}] \leq c^{1/2}\sigma + o(1) \tag{3.7.3}$$

and

$$[E(\boldsymbol{\phi}'\mathbf{H}_j\boldsymbol{\phi})^2]^{1/2} \leq c^{1/2}\sigma + o(1) \tag{3.7.4}$$

with probability one as $n \to \infty$.

Proof Combining NRC and (3.2.10) we find $\Delta_j = E[\boldsymbol{\phi}'\mathbf{H}_j(\boldsymbol{\theta})\boldsymbol{\phi}/2]^2/\sigma^4 + o(n^{-1/2})$. Clearly, $\Delta_j \leq c/\sigma^2$ implies (3.7.4). We derive (3.7.3) using Jensen's inequality, namely $E[(\boldsymbol{\phi}'\mathbf{H}_j\boldsymbol{\phi})^2] \geq [E(\boldsymbol{\phi}'\mathbf{H}_j\boldsymbol{\phi})]^2$.

The previous theorem spells out the conditions under which a proposed model function can be linearized. Substituting $\hat{\Delta}_j$, given by (3.4.7), and consistent variance estimate $\hat{\sigma}^2$, we observe that, in the limit, $\sigma^2\Delta_j \leq c$ implies that the mean and standard deviation of $\boldsymbol{\phi}'\mathbf{H}_j\boldsymbol{\phi}/2$ is bounded above by $c^{1/2}\sigma$. For $f_i \in \mathcal{S}$ the proposed nonlinear assessment statistic at location $\mathbf{x}_j$ is

$$\mathrm{NF}[\hat{f}_j(i)] = \hat{\sigma}\hat{\Delta}_j^{1/2} \tag{3.7.5}$$

for $j \geq 1$. Confidence regions for NF can be constructed using Monte Carlo techniques and is left to Chapter 4. Similar to DFs for stability and accuracy, (3.7.5) can be used to locate regions of the design space, as designated by small NF values, where estimated model functions may be linearized.

Remark 3.7.1

Overall measures of nonlinearity have appeared in the literature, the first being that of Beale (1960). In a paper on nonlinear estimation, Box (1971) demonstrates a connection between Beale's criterion and $\hat{\sigma}^2\hat{\Delta}_j$. Box proposes smallness of

$$\mathrm{NF}_B(\hat{f}_i) = \hat{\sigma}^2 \sum_{j=1}^{n} \hat{\Delta}_j/(m+2) \tag{3.7.6}$$

as a prerequisite for applying linear statistical procedures on the fitted model function or estimated parameters.

In applying the linear approximation (3.7.2), convergence of the matrix of first-order partial derivatives, $\hat{\mathbf{F}}_j$, needs to be assumed. In addition to resampling or simulation measures for checking this condition, the statistic $\hat{\sigma}^2\,\mathrm{Tr}[\hat{\mathbf{H}}_j\hat{\mathbf{M}}\hat{\mathbf{H}}_j]$ for $1 \leq j \leq n$ is proposed for this assessment (see Problem 3.11).

So far we have examined the extent to which an estimated model function can be treated as "linear" in fitted parameters. The magnitude to which the LS estimator departs from linearity has been analyzed using the mathematical notion of curvature. Bates and Watts (1980) describe two fundamental measures of nonlinearity based on functions of the first- and second-order partial derivatives of the model function. For a proposed model from S let $\mathbf{V}$ be the $n \times m \times m$ array consisting of elements $f_{jrs} = \partial^2 f_j(\boldsymbol{\theta}_o)/(\partial\theta_r\partial\theta_s)$ and $\rho = [m\sigma_f^2]^{1/2}$, called the standard radius. For $n \times 1$ vector $\mathbf{h}$ define $\dot{W}_{\mathbf{h}} = \mathbf{Fh}/\rho$ and $\ddot{W}_{\mathbf{h}}$ as an $n \times 1$ vector consisting of elements $\mathbf{h}'\mathbf{H}_j\mathbf{h}$ for $1 \leq j \leq n$.

For nonlinear analysis $\ddot{W}_{\mathbf{h}}$ is decomposed into two portions: $\mathbf{W}_{\mathbf{h}}^{\mathrm{IN}}$, the intrinsic part, which is independent of the form of the parameter, and the parameter effects part $\mathbf{W}_{\mathbf{h}}^{\mathrm{PE}}$. The intrinsic and parameter effects curvatures are defined by

$$\overset{\mathrm{IN}}{\underset{\mathbf{h}}{\mathrm{K}}} = \|\mathbf{W}_{\mathbf{h}}^{\mathrm{IN}}\|/\|\dot{W}_{\mathbf{h}}\|^2 \tag{3.7.7}$$

and

$$\overset{\mathrm{PE}}{\underset{\mathbf{h}}{\mathrm{K}}} = \|\mathbf{W}_{\mathbf{h}}^{\mathrm{PE}}\|/\|\dot{W}_{\mathbf{h}}\|^2 \tag{3.7.8}$$

For a prospective model, taking the supremum of (3.7.6) and (3.7.7) with respect to unit vectors **h** yields the intrinsic curvature IN and the parameter effects curvature PE, respectively. The procedure for finding IN and PE is expanded on in Problem 3.12. A statistical test of significance of IN and PE, at significance level α, compares these values with $1/(2F^{1/2})$ where $F = F_{(1-\alpha,m,n-m)}$. Hence, if PE and IN are less than $1/(2F^{1/2}_{(1-\alpha,m,n-m)})$ then $\hat{\boldsymbol{\theta}}$ is considered to be linear in nature (approximated by (3.2.10)).

The importance of these curvature measures is brought out by the fact that PE, unlike IN, is affected by the form of the model function. Effective transformations or reparametrizations yielding linearizable model functions can be sought after using the PE statistic. Ratkowsky (1983) fully investigates this procedure in choosing judicious reparametrizations. For linearizable model functions, linear statistical inferences such as confidence regions and hypothesis tests on estimators can be constructed (see Section 3.9).

Although the sampling properties of the fitted model functions are unaffected by reparametrizations, the role of transformations in obtaining "normal" LS estimators is still important. The DFs proposed in this chapter, namely SF, AF, and CF, depend on linear and quadratic Taylor series approximations. To this end, model functions whose parameter estimates are close to linear in their sampling properties produce more accurate approximations. This is demonstrated in the following example.

Example 3.7.1

The magnitude of the nonlinearity of the fitted model functions introduced in Example 3.4.2 is examined. To demonstrate the usefulness of reparametrizations we also consider an alternative form of f_1, denoted f_4, given by

$$f_j(4, \boldsymbol{\theta}) = \exp(\theta_1 - \theta_2 \theta_3^{x_j})$$

for $\boldsymbol{\theta}' = (\theta_1, \theta_2, \theta_3)$. The LS estimators are $\hat{\boldsymbol{\theta}}_1 = 4.417$, $\hat{\boldsymbol{\theta}}_2 = 3.401$, and $\hat{\boldsymbol{\theta}}_3 = .964$. Measures of nonlinearity, namely IN, PE, and NF_B, using $\hat{\sigma}^2 = 1.3$ for the models are listed in Table 3.10. As noted in Example 3.6.1, we see from Table 3.10 that model f_3 contains the most nonlinearity while f_2 is the closest to linear in nature. Also, the reparametrization reduced PE from 2.324 for f_1 to 1.644 for f_4. To further explore the effects of transformation, $\hat{\Delta}$ and CF for

Table 3.10 Nonlinearity Statistic

	Statistic NL_B			
Values	f_1	f_2	f_3	f_4
IN	.09	.073	.267	.09
PE	2.324	1.644	6.679	1.644
NF	.097	.021	1.56	.023

Table 3.11 Reparametrizations

		Statistics		
		Δ	CF	NF_B
Original	f_1	.374	15.865	.097
Reparametrized	f_4	.090	14.428	.023

f_4 are found. These values along with NF_B for f_4 are given in Table 3.11.

Based on Table 3.11, the effects of reformulating model f_1 as f_4 are as expected. We find all statistics CF and NL_B, decrease in going from f_1 to f_4. The functional form of f_1 results in an inflation of 1.437 in CF over that of f_4. Here we see the necessity for efficient reparametrization.

Nonlinear DFs, such as NF and NF_B, can be used when searching for an efficient model function formulation. These values can easily be calculated for various forms of the model function and the one associated with the least nonlinearity measures can be utilized. Measures of the fitted model's linearizability, as a function of $\hat{\Delta}$, are related curvature measures IN and PE. Thus, a reduction in NF_B will correspond to a reduction in PE.

The ability to reparametrize the model function, thereby reducing the amount of nonlinearity as measured by $\hat{\Delta}$ or PE, leads to more accuracy when solving for LS estimators using iterative procedures. This is true for the Gauss–Newton method as defined in (3.2.6). Other methods, some of which are modifications of this method, are given in Bard (1974) and Kennedy and Gentle (1980). Also, good initial parameter estimates, in the neighborhood of the true parameter, may be required for convergence properties. This idea is utilized when simulation methods are explored in Chapter 4, where convergence, and sometimes quick convergence, for generated data sets is necessary.

3.8 TESTS ON LINEAR MODELS

The most common situation for model discrimination problems occurs when linear, in parameter, model functions are contrasted. These models frequently occur in the data analysis setting. In this section we take the view of Atkinson's A1 or A3, assuming a true model exists among the candidates, and construct DMs which allow approximate selection probabilities or bounds. Based on these probabilities, formal test procedures are constructed. Unlike the settings of Sections 3.5 and 3.6, more common in the social sciences than in engineering or business, where no model is taken as the true one, more structure and rigorous mathematical techniques are required here.

The competing model functions of $\mathcal{S}$ are assumed to be linear in parameters or, at least, linearizable (see Section 3.7). Since in the general linear regression model function an intercept term is present, one of the predictor variables is taken to be identically one for all locations. The modeling is accomplished by regressing response variable Y_j on the vector of independent variables $\mathbf{x}_j = (1, x_{j1}, \ldots, x_{j(m-1)})$ for $j = 1, \ldots, n$. Since differing model functions, and functions of these model functions, are contrasted, the indexing of the specific models is done throughout this section. For $f_i \in \mathcal{S}$ the regression modeling scheme (3.2.1) becomes

$$Y_j = \mathbf{x}_j(i)\boldsymbol{\beta}(i) + e_j(i) \tag{3.8.1}$$

where, as in Example 1.2.1, $\mathbf{x}_j(i)$ is the $1 \times m(i)$ vector of design location values and $\boldsymbol{\beta}(i)$ is the $m(i) \times 1$ vector of unknown parameters. Under the true model taking parameter $\beta = \beta_0 \in R^{m(i)}$ we assume $e_j(i) \sim$ i.i.d.$n(0, \sigma^2)$. For linearizable model functions consisting of $m(i) - 1$ predictor variables $\mathbf{x}_j(i) = (1, \mathbf{F}_j(i)')$ for $j \geq 1$. The $n \times m(i)$ design matrix is denoted $\mathbf{x}(i)$ and we define the idempotent matrix $\mathbf{E}(i) = \mathbf{x}(i)[\mathbf{x}(i)'\mathbf{x}(i)]^{-1}\mathbf{x}(i)'$. Letting the vectors of responses and errors for f_i be $\mathbf{Y}$ and $\mathbf{e}(i)$, (3.8.1) becomes

$$\mathbf{Y} = \mathbf{x}(i)\boldsymbol{\beta}(i) + \mathbf{e}(i) \tag{3.8.2}$$

The LS estimator of $\boldsymbol{\beta}(i)$ is $\hat{\boldsymbol{\beta}}(i) = [\mathbf{x}(i)'\mathbf{x}(i)]^{-1}\mathbf{x}(i)'\mathbf{Y}$ with fitted model function values given in the vector $\hat{\mathbf{f}}(i) = \mathbf{E}(i)\mathbf{Y}$. Further, the vector of residuals, for model f_i, is $\mathbf{R}(i) = [\mathbf{I} - \mathbf{E}(i)]\mathbf{Y}$.

In the case of nested linear model functions F tests, where the smaller model is assumed under H_0, have been derived and are given by (1.5.10). The case of general linear regression model testing is now explored. For two linear model functions f_r and f_i from $\mathcal{S}$, consider the class of LQMs given by C_1 which employ the common choices of $\mathbf{S}$ for $\mathbf{S}(q)$ and $c(q) = 0$, for $q = r, i$, in (1.4.3). The calculations of conditional IS probabilities are more complex because of the dependence of the fitted model function values on the responses. In fact, if f_r is selected over f_i the dimensions of the residuals $\mathbf{R}(r)$ and estimated model function differences $\hat{\mathbf{f}}(r,i) = \hat{\mathbf{f}}(r) - \hat{\mathbf{f}}(i) = [\mathbf{E}(r) - \mathbf{E}(i)]\mathbf{Y}$ must be constricted. This is accomplished using the $n \times s$ matrix $\mathbf{P}$ where $s \leq [n/2]$ and

$$\begin{pmatrix} \mathbf{P}'[\mathbf{E}(r) - \mathbf{E}(i)] \\ \mathbf{P}'[\mathbf{I} - \mathbf{E}(i)] \end{pmatrix} \quad \text{has full row rank} \tag{3.8.3}$$

For ease of notation let $\mathbf{R} = \mathbf{P}'\mathbf{R}(i)$ and $\mathbf{D} = \mathbf{P}'\hat{\mathbf{f}}(r,i)$. Further, noting the connection between C_1 and C_2 demonstrated in Theorem 2.2.1, we consider LQM in C_2 where $\mathbf{A} = \mathbf{PB}$ for $n \times 1$ vector $\mathbf{B}$. From (1.6.1) and the proof of Theorem 2.2.1 we find that f_r is selected over f_i if

$$2\mathbf{B}'\mathbf{R} < \mathbf{B}'\mathbf{D} \qquad \text{if } \mathbf{B}'\mathbf{D} < 0 \qquad \text{and} \qquad 2\mathbf{B}'\mathbf{R} > \mathbf{B}'\mathbf{D} \qquad \text{if } \mathbf{B}'\mathbf{D} \geq 0 \tag{3.8.4}$$

The ramifications of condition (3.8.3) are explored in the following lemma.

Lemma 3.8.1

Let (3.8.3) hold where the true model is f_i, that is $E(\mathbf{Y}) = \mathbf{x}(i)\boldsymbol{\beta}(i) = \boldsymbol{\mu}$. Let $\mathbf{Y}$ be multivariate normal with dispersion matrix $\sigma^2\mathbf{I}$. Then

$$D \sim N(\mathbf{P}'[\mathbf{E}(1) - \mathbf{E}(2)]\boldsymbol{\mu}, \sigma^2\mathbf{P}'\mathbf{G}\mathbf{P}) \tag{3.8.5}$$

where $\mathbf{G} = \mathbf{E}(r) + \mathbf{E}(i) - \mathbf{E}(r)\mathbf{E}(i) - \mathbf{E}(i)\mathbf{E}(r)$. In addition, the conditional distribution of $\mathbf{R}$ given $\mathbf{D} = \mathbf{d}$ is

$$R \mid D = d \sim N(\boldsymbol{\mu}_c, \sigma^2\Sigma_c) \tag{3.8.6}$$

where

$$\boldsymbol{\mu}_c = \mathbf{P}'[\mathbf{I} - \mathbf{E}(i)]\mathbf{E}(r)\mathbf{P}\mathbf{W}\{\mathbf{d} - \mathbf{P}'[\mathbf{E}(r) - \mathbf{E}(i)]\boldsymbol{\mu}\} \tag{3.8.7}$$

for $\mathbf{W} = (\mathbf{P}'\mathbf{G}\mathbf{P})^{-1}$ and

$$\begin{aligned}\Sigma_c = {} & \mathbf{P}'[\mathbf{I} - \mathbf{E}(i)]\mathbf{P} - \mathbf{P}'[\mathbf{I} - \mathbf{E}(i)]\mathbf{E}(r)\mathbf{P}\mathbf{W} \\ & \times \mathbf{P}'\mathbf{E}(r)[\mathbf{I} - \mathbf{E}(i)]\mathbf{P}\end{aligned} \tag{3.8.8}$$

Proof The proof is left as an exercise.

Choosing $\mathbf{P}$ such that (3.8.3) holds defines a subclass of LQMs of C_2 when f_r is selected over f_i. Let this class be $C_{r|i}(\mathbf{P})$ = LQMs of C_2 where $\mathbf{A} = \mathbf{PB}$. The conditional IS probability is derived in the following theorem.

Theorem 3.8.1

Under the assumptions of Lemma 3.8.1 using an LQM in $C_{r|i}(\mathbf{P})$

$$P(f_r \mid f_i) = \int P\left\{Z > |\mathbf{B}'\mathbf{d}|/[2(\mathbf{B}'\Sigma_c\mathbf{B})^{1/2}] - \delta\right\} dF(\mathbf{D} = \mathbf{d}) \quad (3.8.9)$$

where

$$\delta = \mathbf{B}'\mathbf{d}\mathbf{B}'\boldsymbol{\mu}_c/[\sigma|\mathbf{B}'\mathbf{d}|(\mathbf{B}'\Sigma_c\mathbf{B})^{1/2}] \quad (3.8.10)$$

and the distribution of $\mathbf{D}$ is given by (3.8.5).

Proof The proof is completed using Lemma 3.8.1 and (3.8.4).

From the general form (3.8.9) we note that the calculation of conditional IS probabilities is difficult at best. A subclass of DMs of $C_{r|i}(\mathbf{P})$ which allow IS probability or upper bound calculation is constructed. Let a collection of eigenvectors associated with zero eigenvalues of $\mathbf{E}(r)$ be $\mathbf{V}_1, \ldots, \mathbf{V}_s$ for $s \le [n/2]$. Define $\mathbf{P}(r) = (\mathbf{V}_r, \ldots, \mathbf{V}_s)$ such that condition (3.8.3) is assumed to hold. For such a $\mathbf{P}$ we denote the class of DM in $C_{r|i}(\mathbf{P})$, whose elements depend on the choice of $\mathbf{B}$, by $C_{r|i}(\mathbf{P}(r))$. Our attention is turned to finding an optimal procedure in $C_{r|i}(\mathbf{P}(r))$.

Theorem 3.8.2

Let $f_i, f_r \in \mathcal{S}$ where normality of the error term holds. Using an LQM from $C_{r|i}(\mathbf{P}(r))$, δ given by (3.8.10) is zero and

$$P(f_r \mid f_i) = \int P\left\{Z > |\mathbf{B}'\mathbf{d}|/[2\sigma(\mathbf{B}'\Sigma_c\mathbf{B})^{1/2}]\right\} dF(\mathbf{D} = \mathbf{d}) \quad (3.8.11)$$

where $\mathbf{D} = -\mathbf{P}(r)\mathbf{E}(i)\mathbf{Y}$ and $\Sigma_c = \mathbf{P}(r)'[\mathbf{I} - \mathbf{E}(i)]\mathbf{P}(r)$. Further, the LQM in $C_{r|i}(\mathbf{P}(r))$ which takes $\mathbf{B} = [\mathbf{P}(r)'(\mathbf{I} - \mathbf{E}(i))\mathbf{P}(r)]^{-1}\mathbf{D}$ minimizes (3.8.11) with minimum

$$P(f_r \mid f_i) = \int P(\chi^2_{(1)} > q/4)\, dF(Q = q)/2 \qquad (3.8.12)$$

where $Q = \mathbf{Y}'\mathbf{E}(i)\mathbf{P}(r)\Sigma_c^{-1}\mathbf{P}(r)'\mathbf{E}(i)\mathbf{Y}/\sigma^2$.

Proof First, since $\mathbf{E}(r)\mathbf{P} = 0$ we find from (3.8.7) $\boldsymbol{\mu}_c = 0$, which in turn implies δ is zero. Hence (3.8.11) results from (3.8.9). Now from (3.8.11) we see that $P(f_r \mid f_i)$ is minimized if, inside the integral, the $\mathbf{B}$ vector that maximizes $(\mathbf{B}'\mathbf{d})^2/(\mathbf{B}'\Sigma_c\mathbf{B})$ is substituted. Taking the derivative with respect to $\mathbf{B}$ yields solution $\Sigma_c\mathbf{B} = \lambda\mathbf{d}$ for constant λ. Substituting $\mathbf{B} = \Sigma_c^{-1}\mathbf{d}$ into (3.8.11), squaring, and noting that $\mathbf{B} = -\mathbf{P}(r)\mathbf{E}(i)\mathbf{Y}$ completes the proof.

In practice, to calculate the conditional IS probability given in Theorem 3.8.2 we rewrite (3.8.12). Changing the order of integration,

$$P(f_r \mid f_i) = \int_0^\infty P(Q < 4\alpha)(2\pi\alpha)^{-1/2}\exp(-\alpha/2)\,d\alpha \qquad (3.8.13)$$

Further, let $\mathbf{S}$ be the matrix whose columns consist of the eigenvectors of $\mathbf{E}(i)\mathbf{P}(r)\Sigma_c^{-1}\mathbf{P}(r)'\mathbf{E}(i)$ so that Q can be written as the weighted sum of independent noncentral chi-square random variables. Here $Q = \sum \lambda_i W_i^2$ where λ_i and W_i are the positive eigenvalues and associated eigenvectors of $\mathbf{E}(i)\mathbf{P}(r)\Sigma_c^{-1}\mathbf{P}(r)'\mathbf{E}(i)$ and $W_i^2 \sim \chi^{2'}((1, \delta_i)$ where $\delta_i = (\mathbf{S}_i'\boldsymbol{\mu})^2$. The distribution of Q is discussed in the next remark.

Remark 3.8.1

The distribution of a weighted sum of independent noncentral chi-square random variables has been derived by Imhof (1961). For $Q = \sum \lambda_i W_i^2$ where W_i^2 are $\chi^2_{(1,\delta_i)}$ and λ_i are constants for $1 \le i \le s$

$$P(Q > c) = 1/2 + \pi^{-1} \int_O^\infty \{\sin(\theta(u))/[u\phi(u)]\}\, du \qquad (3.8.14)$$

where

$$\theta(u) = .5 \sum_{j=1}^{s} [\tan^{-1}(\lambda_i u) + \delta_i \lambda_i u/(1 + \lambda_i^2 u^2)] - .5cu \qquad (3.8.15)$$

and

$$\phi(u) = \prod_{i=1}^{s} (1 + \lambda_i^2 u^2)^{1/4} \exp(.5 \sum \delta_i \lambda_i^2 \times u^2/(1 + \lambda_i^2 u^2)] \qquad (3.8.16)$$

The conditional IS probability given by (3.8.12) depends on the true parameter value $\boldsymbol{\mu}$ and, hence, is hard to employ in a DM. One procedure is to approximate the integration by resampling (this is left to Chapter 4). Another method is to employ Remark 3.8.1 and the fact that (3.8.15) is maximized when $\boldsymbol{\mu} = 0$. In this setting we have the upper bound

$$P(f_r \mid f_i) \le 1/2 - (2\pi^3)^{-1/2} \int \alpha^{-1/2} \exp(-\alpha/2) \times \sin(\theta(u))/[u\phi(u)]\, du\, d\alpha \qquad (3.8.17)$$

where, putting $\delta_i = 0$ for $i \ge 1$ in (3.8.15) and (3.8.16),

$$\theta(u) = \left[\sum_{i=1}^{s} \tan^{-1}(\lambda_i u) - \alpha u\right] .5 \qquad (3.8.18)$$

and

$$\phi(u) = \prod_{i=1}^{s}(1 + \lambda_i^2 u^2)^{1/4} \tag{3.8.19}$$

The details of the numerical integration in this bound are left to the reader in Problem 3.24.

Another method of estimating (3.8.12), in the case of large samples, results if Q acts as a constant. From the theory of the distribution of quadratic forms

$$\begin{aligned}\text{Var}(Q) = {} & 2\,\text{Tr}[\mathbf{E}(i)\mathbf{P}\Sigma_c^{-1}\mathbf{P}'\mathbf{E}(i)\mathbf{P}\Sigma_c^{-1}\mathbf{P}'] \\ & + 4\boldsymbol{\mu}'\mathbf{E}(i)\mathbf{P}\Sigma_c^{-1}\mathbf{P}'\mathbf{E}(i)\mathbf{P}\Sigma_c^{-1}\mathbf{P}'\mathbf{E}(i)\boldsymbol{\mu}/\sigma^2\end{aligned} \tag{3.8.20}$$

where, for simplicity, $\mathbf{P} = \mathbf{P}(r)$. It follows that an unbiased estimator, under model correctness, of Var(Q) is, as $n \to \infty$,

$$\begin{aligned}S^2(Q) = {} & 4\mathbf{Y}'\mathbf{E}(i)\mathbf{P}\Sigma_c^{-1}\mathbf{P}'\mathbf{E}(i)\mathbf{P}\Sigma_c^{-1}\mathbf{P}'\mathbf{E}(i)\mathbf{Y}/\hat{\sigma}^2 \\ & - 2\,\text{Tr}[\mathbf{E}(i)\mathbf{P}\Sigma_c^{-1}\mathbf{P}\mathbf{E}(i)\mathbf{P}\Sigma_c^{-1}\mathbf{P}']\end{aligned} \tag{3.8.21}$$

(see Problem 3.15), where $\hat{\sigma}^2$ is a consistent estimator of σ^2. Thus, small $S^2(Q)$ indicates that

$$\hat{q} = \mathbf{y}'\mathbf{E}(i)\mathbf{P}\Sigma_c^{-1}\mathbf{P}'\mathbf{E}(i)\mathbf{y}/\hat{\sigma}^2 \tag{3.8.22}$$

may be substituted for Q in (3.8.2), yielding an approximation for $P(f_r \mid f_i)$,

$$\hat{P}(f_r \mid f_i) = P(\chi^2_{(1)} > \hat{q}/4)/2 \tag{3.8.23}$$

This procedure is demonstrated in the next example.

Example 3.8.1

Two linear models given by (3.8.1) are compared. The first model, without intercept, takes location values $x_j(1) = 1, 2, 2, 3$, while the second model utilizes $x_j(2) = 1, 1, 2, 3$. We let f_2 take the place of f_r in Theorem 3.8.2, forming $C_{r|i}(\mathbf{P}(r))$, and find

$$\mathbf{E}(2) = \begin{pmatrix} .0667 & .0667 & .1333 & .2 \\ .0667 & .0667 & .1333 & .2 \\ .1333 & .1333 & .2667 & .4 \\ .2 & .2 & .4 & .6 \end{pmatrix}$$

For a suitable $\mathbf{P}(2)$ we use

$$\mathbf{P}(2)' = \begin{pmatrix} .7071 & -.7071 & 0 & 0 \\ .3162 & .3162 & .6325 & -.6325 \end{pmatrix}$$

where $\mathbf{P}(2)'\mathbf{E}(2) = \mathbf{0}$. From Theorem 3.8.2 we calculate

$$\Sigma_c = \begin{pmatrix} .9723 & -.0124 \\ -.0124 & .9277 \end{pmatrix}$$

Assuming f_1 is the true model, for selected $\beta(1)$ values Var(Q) is calculated using (3.8.20) and $\hat{P}(f_2 \mid f_1)$, substituting $\boldsymbol{\mu}$ for Y in (3.8.22), is found. To measure the accuracy of $\hat{P}(f_2 \mid f_1)$ given by (3.8.23), a simulation consisting of 1000 runs is conducted in which the LQM in $C_{2|1}(\mathbf{P}(2))$ taking $\mathbf{B} = \Sigma_c^{-1}\mathbf{d}$ is utilized. The observed probability of selecting f_2 over the true model f_1 is found. For comparison, the LSE method is applied to the simulated data and the observed probability of an IS is found. Here, the normally distributed errors have $\sigma^2 = 1/12$ and the results are listed in Table 3.12.

Table 3.12 Linearization Technique

Value	Calculations			Observed	LSE
	$E(Q)$	$\text{Var}(Q)$	$P(f_2 \mid f_1)$	$P(f_2 \mid f_1)$	$P(f_2 \mid f_1)$
.1	.027	.001	.435	.420	.442
.5	.471	.016	.246	.258	.251
1.0	1.858	.064	.086	.084	.044

The small values of $\text{Var}(Q)$ found in Table 3.12 for all selected $\beta(1)$ values indicates that there should be close agreement between the estimate $\hat{P}(f_2 \mid f_1)$ and the true conditional IS probability. This is confirmed by the observed IS probabilities. In addition, for this choice of $\mathbf{P}(2)$ the conditional IS probabilities associated with the LQM compare with that of the LSE method, especially when $\beta(1) = .1$. For other choices $\mathbf{P}(2)$ the resulting DMs possess different sampling characteristics for Q which can be evaluated based on empirical data (see Problem 3.16).

Using Theorem 3.8.2 and applying pairwise DMs described in this section, tests of k competing models can be constructed. Two such DMs, one simultaneous and the other sequential, are now presented.

Example 3.8.2

Testing in Atkinson's A1 setting is considered. A model $f_i \in S$ is fixed and compared against all other models $f_r \in S$. For each $f_r, f_r \neq f_i$, an appropriate $\mathbf{P}(r)$ satisfying (3.8.3) is found and the DM described in Theorem 3.8.2 is applied utilizing the optimal $\mathbf{B} = [\mathbf{P}(r)'(\mathbf{I} - \mathbf{E}(i))\mathbf{P}(r)]^{-1}\mathbf{D}$. Hence, $k - 1$ comparisons are run. Set S is assumed to contain the true model and we test

$$H_0\text{: } f_i \text{ Correct}, f_i \in S$$

against

$$H_1\colon f_r \text{ Correct}, f_r \in \mathcal{S} f_r \neq f_i$$

by rejecting H_0 if f_i is not selected by each pairwise test. Here we have the bound $\alpha \leq \sum P(f_r \mid f_i)$.

Example 3.8.3

In the basic framework of Theorem 2.4.1, an SM is constructed based on Theorem 3.8.2. On each step $f_{(i)} \in \mathcal{S}$ is chosen and a suitable $\mathbf{P}(i)$, satisfying (3.8.3), is found. Pairwise tests of $f_{(i)}$ against its rivals in $\mathcal{S}_i$ are run utilizing the QDF outlined in Theorem 3.8.2 ($\mathbf{B} = [\mathbf{P}(i)'(\mathbf{I} - \mathbf{E}(i))\mathbf{P}(i)]^{-1}\mathbf{D}$) for $i = 1, \ldots, k - 1$. Model $f_{(i)}$ is selected as the true one if it is chosen over each of its competitors at that step. The SM proceeds until a model is selected. As before, conditional IS probabilities can be obtained. Let $p_i = \max P(f_{(i)} \mid f_r)$ where the maximum is taken over all $f_r \in \mathcal{S}_i$. If $f_{(i)}$ is selected

$$P[\text{IS}(f_{(i)})] \leq \max\{p_1, \ldots, p_i\} \tag{3.8.24}$$

The proof of (3.8.24) is left to the reader. We remark that, unlike Theorem 2.4.1, the step decisions of this SM are not independent and the resulting conditional IS probability bounds are not as sharp. Also, the strategy proposed in Section 2.5, selecting "outside" models at each step, seems to be appropriate when there are no prior model beliefs.

The DMs outlined in the previous two examples can become quite involved in their sophistication, and other, simpler methods may be utilized. In the case of two linear model functions another approach is to quantify the observed differences in the model's

QDFs. For f_r and f_i from $\mathcal{S}$ consider the class of LQMs defined by C_1 (common choice of **S** for **S**(s), $s = r,i$, and $c(s) = 0$ for $s = r,i$ in (1.4.3)). The DF of interest is $Q(i,r) = Q(i) - Q(r)$, which can be written as the quadratic form $\mathbf{Y}'\mathbf{B}(i,r)\mathbf{Y}$ where $\mathbf{B}(i,r) = [\mathbf{I} - \mathbf{E}(i)]\mathbf{S}[\mathbf{I} - \mathbf{E}(i)] - [\mathbf{I} - \mathbf{E}(r)]\mathbf{S}[\mathbf{I} - \mathbf{E}(r)]$. With the LQM f_r is selected over f_i if $Q(i,r)$, or $\mathbf{Y}'\mathbf{B}(i,r)\mathbf{Y}$, is positive. The conditional IS probability depends on the true model function and the form of $\mathbf{B}(i,r)$.

For $f_i, f_r \in \mathcal{S}$ let the observed value of $Q(i,r)$ be denoted $q(i,r)$. The magnitude of $q(i,r)$ indicates the strength of the model selection. In the development that follows the LQM is employed and we search for efficient and obtainable bounds on

$$P_{r|i}(c) = P[Q(i,r) \geq c \mid f_i \text{ True}] \qquad (3.8.25)$$

for constant c. Thus, letting $c = q(i,r)$, the IS probability (3.8.25) or appropriate upper bound can be used to construct tests under A1 or A3. Such a probability statement is given in the following theorem.

Theorem 3.8.3

Under models f_i and f_r given by (3.8.1) let $\mathbf{I} - \mathbf{E}(i) - \mathbf{B}(i,r)$ be nonnegative definite for $i,r = 1,2$ that is, let

$$\mathbf{Y}'[\mathbf{I} - \mathbf{E}(i) - \mathbf{B}(i,r)]\mathbf{Y} \geq 0 \qquad (3.8.26)$$

with probability one. Under normality of the error term

$$P_{r|i}(c) = P(\chi^2_{n-m(i))} \geq c/\sigma^2)\Delta(c,r,i) \qquad (3.8.27)$$

where

$$\Delta(c,r,i) = 1 - P[Q(i,r) < c \mid \mathrm{RSS}(f_i) \geq c)] \tag{3.8.28}$$

for $i \neq r$.

Proof Using (3.8.26) we find

$$P[\mathbf{Y}'\mathbf{B}(i,r)\mathbf{Y} \geq c] = P[\mathbf{Y}'(I - \mathbf{E}(i))\mathbf{Y} \geq c) \\ - P[\mathbf{Y}'\mathbf{B}(i,r)\mathbf{Y} < c \leq \mathbf{Y}'(I - \mathbf{E}(i))\mathbf{Y}]$$

The result, (3.8.27), is obtained noting that under the true model f_i, $\mathbf{Y}'(\mathbf{I} - \mathbf{E}(i))\mathbf{Y}/\sigma^2 \sim \chi^2_{n-m(i))}$.

The IS probability (3.8.30) has the advantage of ease of calculation over the methods of $C_{r|i}(\mathbf{P}(r))$. However, in any particular case the value of $\Delta(c,r,i)$ must be estimated. This can be done using resampling methods introduced in Chapter 4. It is left to the reader to construct tests of A1, similar to Example 3.8.2 and SMs analogous to Example 3.8.3, utilizing Theorem 3.8.3. We also note that condition (3.8.26) is satisfied in all cases by $\mathbf{S} = \mathbf{I}$ or $\mathbf{I} - \mathbf{E}(1) - \mathbf{E}(2)$ (see Problem 3.17).

Remark 3.8.2

The analysis of IS probabilities resulting from LQMs was performed by Borowiak (1981). With the additional conditions:

The nonzero eigenvalues of $\mathbf{B}(i,r)$ occur in plus-minus pairs; that is, the nonzero eigenvalues are $\pm\lambda_1, \ldots, \pm\lambda_s$ where $s \leq n/2$ (3.8.29)

and

$$\text{If } f_i \text{ is the true model, then } E[\mathbf{Y}'\mathbf{B}(i,r)\mathbf{Y}] \leq 0 \text{ for } r \neq i \quad (3.8.30)$$

IS probability (3.8.27) is improved to

$$P_{r|i}(c) \leq \min\{1/2, P[\chi^2_{n-m(i))} > c/\sigma^2]\Delta(c,r,i) \quad (3.8.31)$$

The application of (3.8.31) is considered in Problem 3.17. In the original work the value $\Delta(c,r,i) = 1/2$ was proposed on the basis of empirical results.

Condition (3.8.30) implies, in an intuitive manner, that the LQM is fair under each possible model. The following example dealing with time series models demonstrates testing nonnested model functions.

Example 3.8.4

Propane gas bills over a 7-year stretch are monitored quarterly. In modeling these bills two effects, the first due to a seasonal adjustment and the second due to an increasing trend, must be taken into account. Two time series linear regression models are developed. Let x_j denote the order of the response from period to period; thus $x_j = 1, 2, \ldots, 28$. The first model function takes the form

$$f_j(1) = \beta_0(1) + \beta_1(1)x_j + \beta_2(1)\sin(2\pi x_j/L) + \beta_3(1)\cos(2\pi x_j/L)$$

where L is the number of periods in a year ($L = 4$). The second model function is

$$f_j(2) = \beta_0(2) + \beta_1(2)x_j + \beta_2(2)x_{1J} + \beta_3(2)x_{2j} + \beta_4(2)x_{3j}$$

where the dummy variables defined by

$$x_{rj} = \begin{cases} 1 & \text{if } x_j \text{ is in the } r\text{th quarter} \\ 0 & \text{otherwise} \end{cases}$$

for $r = 1, 2, 3$ model the seasonal effect. The design matrices corresponding to f_1 and f_2 take the forms

$$\mathbf{x}(1) = \begin{pmatrix} 1 & 1 & 1 & 0 \\ 1 & 2 & 0 & -1 \\ 1 & 3 & -1 & 0 \\ 1 & 4 & 0 & 1 \\ \vdots & \vdots & \vdots & \vdots \end{pmatrix}$$

and

$$\mathbf{x}(2) = \begin{pmatrix} 1 & 1 & 1 & 0 & 0 \\ 1 & 2 & 0 & 1 & 0 \\ 1 & 3 & 0 & 0 & 1 \\ 1 & 4 & 0 & 0 & 0 \\ \vdots & \vdots & \vdots & \vdots & \vdots \end{pmatrix}$$

We note that f_1 is nested in f_2 (see Problem 3.19) and, since $m(i)$ denotes the number of independent variables minus one, $m(1) = 4$ while $m(2) = 5$.

The responses, consisting of 28 bills taken from Bowerman et al. (1986), are listed in Table 3.13. The LS estimators are found along with RSS and $S^2(\hat{f}) = [\mathrm{RSS}(\hat{f}_i)/(n - m(i))]^{1/2}$ and are given in Table 3.14. Model f_1, because of its one less predictor variable, has a better fit than f_2, as measured by $S^2(\hat{f})$. To test this model rigor-

Table 3.13 Propane Gas Data

Year	Quarter 1	Quarter 2	Quarter 3	Quarter 4
1	344.39	246.63	131.53	288.87
2	313.45	189.76	179.10	221.10
3	246.84	109.00	51.21	133.89
4	277.01	197.98	50.68	218.08
5	365.10	207.51	54.63	214.09
6	267.00	230.28	230.32	426.41
7	467.06	306.03	253.23	279.46

ously we need to test hypothesis

$$H_0\colon f_2 \text{ Correct}$$

against

$$H_1\colon f_1 \text{ Incorrect} \tag{3.8.32}$$

Since the standard test of nested linear model functions, given by (1.5.10), takes f_1 under the null hypothesis, a DM from this section is required.

To test (3.8.32) we note that only one eigenvector (not unique) corresponding to a zero eigenvalue of $\mathbf{E}(1)$, and not $E(2)$, exists. A suitable $n \times 1$ vector $\mathbf{P}(1)$, such that $\mathbf{P}(1)'\mathbf{x}(1) = \mathbf{0}$, is listed in Table 3.14. Using Theorem 3.8.2, $D = -\mathbf{P}(1)'\mathbf{x}(2)\boldsymbol{\beta}(2) = 171.366$ and $\Sigma_c = 3.348$, which in turn yields $B = 51.182$. Hence, $BD > 0$

Table 3.14 Time Series Regression Models

Model	LS estimators					RSS	S^2
f_1	168.34	8.257	101.05	26.32		102,751	65.3
f_2	202.67	8.202	60.336	−52.75	−141.66	101,328	66.3

and condition (3.8.4) is satisfied with $16,629.59 > 8,770.77$. Therefore, utilizing the DM outlined in Theorem 3.8.2, model f_1 is selected. Further, the observed value of Q is found to be $q = 2.0135$. To check the applicability of the conditional IS probability (3.8.23), $S_2(Q)$ given by (3.8.21) is calculated, which is found to be slightly negative. Thus, we consider the variance of Q, (3.8.20), to be small and use (3.8.23) as an estimate of $P(f_1 \mid f_2)$. So $\hat{P}(f_1 \mid f_2) = P(\chi^2_{(1)} > 2.0135/4)/2 = .239$. In the view of hypothesis testing, H_0, listed in (3.8.32), is rejected with significance level $\alpha = \hat{P}(f_1 \mid f_2) = .239$. Although the level of this test is large (because of the closeness of the competing models), this DM yields a test procedure in the nonnested setting.

This section on linear models concludes with an examination of the backwards elimination technique. Such SMs are common to many applications. Assume $\mathcal{S}$ consists of competing nested linear regression models, $f_1 \subset f_2 \subset \cdots \subset f_k$, is assumed to contain the true model, and for convenience $m(i) - m(i-1) = 1$ for $i = 2, \ldots, k-1$. The backward elimination SM utilizes, on each step, QDFs of the form (1.4.3) with $\mathbf{S}(i) = \mathbf{I}$ for $1 \le i \le k$. The SM proceeds from step to step comparing $Q(i)$ to $Q(i-1)$ with $Q(i-1,i) = Q(i-1) - Q(i)$ and stops and declares f_i the true model the first time $Q(i-1,i) > 0$. This SM is denoted SM_B. The IS probabilities and bounds are derived with the aid of the following lemma.

Lemma 3.8.2

For $f_1 \subset f_2 \subset \cdots \subset f_k$ and $m(i) - m(i-1) = 1$, for $i \ge 2$, if $\mathbf{B}(i-1,i) = \mathbf{E}(i) - \mathbf{E}(i-1)$ then $\mathbf{B}(i-1,i)\mathbf{B}(r-1,r) = \mathbf{0}$ for $1 < r < i \le k$. Further, for $i \ge 2$, $\mathbf{B}(i-1,i)$ is idempotent of rank one.

Proof The proof is left to the reader.

Letting $c_r - c_{r-1} = c_{r,r-1}$ the step criterion $Q(i-1,i) > 0$ is $\mathbf{Y}'\mathbf{B}(i-1,i)\mathbf{Y} > c_{i,i-1}$. Lemma 3.8.2 implies that the step decisions are independent. Using the normality of the responses and Lemma 3.8.2 for $r < k$,

$$P(f_k \mid f_r) = P(\chi^2_{(1)} \le c_{k,k-1}/\sigma^2) \tag{3.8.33}$$

Also, for $i < r < k$

$$P(f_i \mid f_r) \le \prod_{j=r}^{k} P(\chi^2_{(1)} \le c_{j,j-1}/\sigma^2) \tag{3.8.34}$$

and for $r < i < k$

$$P(f_i \mid f_r) = \prod_{j=i+1}^{k} P(\chi^2_{(1)} \le c_{j,j-1}/\sigma^2) \times P(\chi^2_{(1)} > c_{i,i-1}/\sigma^2) \tag{3.8.35}$$

The inequality in (3.8.34) is due to the rth step decision.

In any backward SM, larger models or models containing many unknown parameters are favored. For this reason overfitting the data or selecting f_i when, in fact, f_r is correct where $r < i$ is the major problem. The conditional probability of overfitting the data using SM_B, denoted $\mathrm{OF}(\hat{f}_i)$, is given by (3.8.33) and (3.8.35). Also, the probability of an overfit prior to model selection is denoted OF and has probability given by

$$P(\mathrm{OF}) = \max\left\{p_k, (1-p_k)p_{k-1}, \ldots, \prod_{j=2}^{k}(1-p_j)p_1\right\} \tag{3.8.36}$$

where

$$p_j = P(\chi^2 > c_{j,j-1}/\sigma^2) \tag{3.8.37}$$

This probability of an OF can be minimized and is outlined in the next theorem.

Theorem 3.8.4

Under normality of the responses SM_B is employed with $c_{r,r-1} = \sigma^2\chi^2_{1-1/r,1)}$. For this SM_B (3.8.36) is minimized and $P(\text{OF}) = 1/k$.

Proof The minimum of (3.8.36) is attained when $p = (1 - p_k)p_{k-1} = \cdots = (1 - p_j)p_1$. This leads to $p_i = 1/i$, for $i \geq 1$, and the proof is completed using (3.8.37).

The case where σ^2 is estimated by an independent, of model's residuals, source such as $\hat{\sigma}^2$ given by (2.2.6) is considered in the next theorem.

Theorem 3.8.5

Under normality of the responses let $\hat{\sigma}^2$ be independent of $Q(i)$, $1 \leq i \leq k$, and $\hat{\sigma}^2/\sigma^2 \sim \chi^2$ with degrees of freedom m. The SM_B utilizing $c_{r,r-1} = F_{1-1/r,1,m)}\hat{\sigma}^2$ for $2 \leq r \leq k$ attains the minimum $P(\text{OF}) = 1/k$.

Proof The proof follows along lines similar to that of Theorem 3.8.4.

The previous theorem defines an optimal, in terms of OF probabilities, backward SM. This section concludes with an example demonstrating this procedure.

Example 3.8.5

The SM_B procedure is applied to the Hald data introduced in Example 3.6.2. Based on the CF_L DF given by (3.6.17), a collection of nested models, defined by the predictor variables they contain, is chosen (see Table 3.9). They are f_1 employing x_2, f_2 containing x_1, x_2, f_3 consisting of x_1, x_2, x_3, and the full model f_4. To apply the results of Theorem 3.8.5 we look for repeat or "near" repeat data points from which to compute $\hat{\sigma}^2$. Data points with almost identical predictor variable values occur at two locations. At one the responses are 104.3 and 109.2, while at the other we find 113.3 and 109.4. Using (2.2.6) we calculate $\hat{\sigma}^2 = 9.805$ where $m = 2$.

The SM_B procedure is run utilizing the "cutoffs" proposed in Theorem 3.8.5. On each step we find

$$Q(i-1,i) = \text{RSS}(\hat{f}_{i-1}) - \text{RSS}(\hat{f}_i)$$

The required values of the F statistic are $F_{(.75,1,2)} = 2.579$, $F_{(.666,1,2)} = 1.6$, and $F_{(.5,1,2)}$, which lead to the specified $c_{r,r-1}$, namely $c_{r,r-1} = F_{(1-1/r,1,2)}\hat{\sigma}^2$. Table 3.15 lists for each model RSS and for each step (under the larger model) the $Q(i-1,i)$ and $c_{i,i-1}$. From this table we see that the SM stops on the third step, since $848.43 > 6.57$, and model f_2 is chosen. Here, the probability of OF

Table 3.15 Hald Data—Nested Models

Models	RSS	$Q(i-1,1)$	$C_{j-1,i}$
f_4	47.863	.247	25.29
f_3	48.110	9.794	15.69
f_2	57.904	848.43	6.57
f_1	906.336	—	—

is .25. This result corroborates the minimum CF_L criterion, which also selected f_2.

3.9 PARAMETER STABILITY

In this chapter we have, so far, been concerned with the estimated model function's two basic properties, fit to the responses and stability. It sometimes occurs in modeling that one or more of the unknown parameters are of primary interest. Similar to the stability of a fitted model function, the estimated parameters depend on the particular observed set of responses. The closeness of the LS estimators for a proposed model function to the true parametric values in terms of its sampling properties is referred to as parameter stability. In this section DFs which measure the proposed model's parameter stability are constructed. These can be used in conjunction with Atkinson's A2 to evaluate prospective models in the nonhypothesis testing situation.

Based on the large sample properties of the LS estimator, confidence regions for the true unknown parameters can be constructed, and their size indicates the accuracy at which the parameters are estimated. For a fixed model from $\mathcal{S}$ using the limiting distribution of $\hat{\boldsymbol{\theta}}$ given in (3.2.5), an asymptotic $1-\alpha$ confidence region for $\hat{\boldsymbol{\theta}}$ is

$$(\hat{\boldsymbol{\theta}}-\boldsymbol{\theta})'\mathrm{F}'\mathrm{F}(\hat{\boldsymbol{\theta}}-\boldsymbol{\theta}) \leq m\hat{\sigma}^2 F \tag{3.9.1}$$

where $\hat{\sigma}^2$ is a consistent estimator of σ^2 with ν degrees of freedom and $F = F_{(1-\alpha,m,\nu)}$. The accuracy of this large sample region depends on the linear sampling nature of $\hat{\theta}$ and can be checked by simulation resampling.

The rival models in $\mathcal{S}$ are assumed to contain all the parameters of interest. Thus, by rearrangement the proposed models compete on the set of unknown parameters $\theta_1,\ldots,\theta_s$ where $s \leq m$ for all $i = 1,\ldots,k$. Since the various models in $\mathcal{S}$ contain differing m values, some adjustment in (3.9.1) is required. For a proposed model let $\boldsymbol{\theta}' = (\boldsymbol{\theta}_1', \boldsymbol{\theta}_2')$ where $\boldsymbol{\theta}_1' = (\theta_1,\ldots,\theta_s)$ and $\boldsymbol{\theta}_2' = (\theta_{s+1},\ldots,\theta_m)$. The partitions associated with that of $\hat{\boldsymbol{\theta}}$ are $\hat{\boldsymbol{\theta}}' = (\hat{\boldsymbol{\theta}}_1', \hat{\boldsymbol{\theta}}_2')$ and

$$\mathbf{M} = \begin{pmatrix} \mathbf{M}_{11} & \mathbf{M}_{12} \\ \mathbf{M}_{21} & \mathbf{M}_{22} \end{pmatrix}$$

From (3.2.5) eliminating the nuisance parameter $\boldsymbol{\theta}_2$ yields the marginal distribution

$$\hat{\boldsymbol{\theta}}_1 \sim N(\boldsymbol{\theta}_1, \sigma^2 M_{11}) \tag{3.9.2}$$

Therefore, the analog of the confidence region given in (3.9.1) for $\boldsymbol{\theta}_1$ is

$$(\hat{\boldsymbol{\theta}}_1 - \boldsymbol{\theta}_1)'\mathbf{M}_{11}^{-1}(\hat{\boldsymbol{\theta}}_1 - \boldsymbol{\theta}_1) \leq s\hat{\sigma}^2 F \tag{3.9.3}$$

To construct a measure of parameter stability we rotate the quadratic form in $\hat{\theta}_1 - \theta_1$ given by $(\hat{\boldsymbol{\theta}}_1 - \boldsymbol{\theta}_1)'\mathbf{M}_{11}^{-1}(\hat{\boldsymbol{\theta}}_1 - \boldsymbol{\theta}_1)$. This is done using an orthogonal matrix $\mathbf{P}$ such that $\mathbf{P}'\mathbf{M}_{11}\mathbf{P} = \Lambda$ where Λ is diagonal consisting of diagonal elements λ_r for $1 \leq r \leq s$. Here the columns of $\mathbf{P}$ are the eigenvectors of $\mathbf{M}_{11}$. With this transformation (3.9.3) becomes

$$\sum_{r=1}^{s} \lambda_r^{-1}\gamma_r^2 \leq s\sigma^2 F_{(1-\alpha,s,\nu)} \tag{3.9.4}$$

where $\gamma = \mathbf{P}(\hat{\boldsymbol{\theta}}_1 - \boldsymbol{\theta}_1) = (\gamma_1,\ldots,\gamma_s)'$.

When comparing model functions in terms of parameter stability the size of the region bounded by the s = dimensional ellipsoid defined by the boundary of (3.9.4) dictates the accuracy of the LS estimators of interest. For a prospective model the s-dimensional volume is proportional to

$$[s\hat{\sigma}^2 F_{(1-\alpha,s,\nu)}]^s \prod_{r=1}^{s} \lambda_r$$

Hence, the DF assessing parameter stability for the model is

$$\mathrm{PF}(f_i) = \prod_{r=1}^{s} \lambda_r \tag{3.9.5}$$

Models corresponding to smaller values of PF possess better estimation, in terms of the sampling properties, of the unknown parameters of interest.

Remark 3.9.1

Parameter stability, as presented in this section, is connected with standard experimental design criteria. In the case of a linear model function with design matrix $\mathbf{x}$ the eigenvalues λ_r, $r = 1, \ldots, m$, correspond to the matrix $\mathbf{M}$. The common forms of design criteria are referred to as D, E, and A optimally and they minimize, respectively, $\prod \lambda_r, \max\{\lambda_1, \ldots, \lambda_m\}$ and $\sum \lambda_r$. Hence, the DF measuring parameters stability, PF, judges competing model functions according to the D optimality criterion. Other DFs for this purpose can be constructed using the other design criteria. For a review of design optimality we refer to Wald (1943), Ehrenfeld (1953), and Kiefer (1975).

Example 3.9.1

Consider the problem of two competing time series linear regression models of Example 3.8.4. After adjusting for seasonal effects we wish to estimate the parameters $\beta_0(i)$ and $\beta_1(i)$ for $i = 1, 2$ in order to predict the increasing trend in propane gas bills. For both models we partition $\boldsymbol{\beta} = \boldsymbol{\theta}$ into components where $\boldsymbol{\beta}_1' = (\beta_0(i), \beta_1(i))$ for $i = 1, 2$. For model f_1

$$M_{11}(1) = \begin{pmatrix} .15258 & -.00806 \\ -.00806 & .00056 \end{pmatrix}$$

with eigenvalues $\lambda_1(1) = .28599$ and $\lambda_2(1) = .00028$, and for model f_2

$$M_{11}(2) = \begin{pmatrix} .28571 & -.00893 \\ -.00893 & .00056 \end{pmatrix}$$

with eigenvalues $\lambda_1(2) = .15301$ and $\lambda_2(2) = .00013$. to measure parameter stability for the proposed models we employ (3.9.5) and find $\mathrm{PF}(\hat{f}_1) = 1.989 \times 10^{-5}$ and $\mathrm{PF}(\hat{f}_2) = 7.9706 \times 10^{-5}$. Hence, model f_1, with fewer total estimated parameters, has the advantage over f_2 in terms of its parameters' stability. This finding is in accordance with the findings of Example 3.8.4.

In the situation where the value of an estimated parameter or set of parameters is the main goal, the LS estimate itself may be adjusted. In fact, for $f_i \in \mathcal{S}$, we can use the criterion function (3.6.1) as the risk function, to be minimized, for estimation. Here the estimator $\hat{\boldsymbol{\theta}}$ of $\boldsymbol{\theta}$ minimizes

$$\mathrm{CE}(\hat{f}_i) = \mathrm{MRSS}(\hat{f}_i) + \lambda J(\hat{\boldsymbol{\theta}}) \tag{3.9.6}$$

for constant λ and penalty function, in terms of parameter stability, $J(\hat{\boldsymbol{\theta}})$. The function $J(\hat{\boldsymbol{\theta}})$ should be such that estimates with large variances produce large $J(\hat{\boldsymbol{\theta}})$ values.

When the model function under investigation is linear or linearizable and $J(\hat{\boldsymbol{\theta}})$ is quadratic in $\hat{\boldsymbol{\theta}}$, that is, $J(\hat{\boldsymbol{\theta}}) = \hat{\boldsymbol{\theta}}'\mathbf{D}\boldsymbol{\theta}$, then the general estimator is

$$\bar{\boldsymbol{\theta}} = \hat{\mathbf{S}}\hat{\mathbf{F}}'\hat{\mathbf{F}}\hat{\boldsymbol{\theta}} \tag{3.9.7}$$

where $\hat{\mathbf{S}} = [\hat{\mathbf{F}}'\hat{\mathbf{F}} + \hat{\mathbf{D}}]^{-1}$ (see Problem 3.20) for symmetric positive definite matrix $\hat{\mathbf{D}}$. Utilization of $\bar{\boldsymbol{\theta}}$ is not a model discrimination problem but a modeling technique. The estimate (3.9.7) is sensitive to the fitted parameters through the choice of $D(i)$. If estimator $\bar{\boldsymbol{\theta}}$ is used, then adjustments to the DFs based on the normal theory and asymptotic expansions are needed. From (3.9.7) under NRC we have the limiting distribution

$$\bar{\boldsymbol{\theta}} \sim N(\mathbf{SF}'\mathbf{F}\boldsymbol{\theta}, \sigma^2\mathbf{SF}'\mathbf{FS}) \tag{3.9.8}$$

and Taylor series analog to (3.2.7) (substituting $\bar{\boldsymbol{\theta}}$ for $\boldsymbol{\theta}$)

$$\hat{f}_j = f_j + \mathbf{F}_j'\mathbf{SF}'\mathbf{F}\boldsymbol{\phi} + \boldsymbol{\phi}'\mathbf{F}'\mathbf{FSH}_j(\boldsymbol{\theta}^*)\mathbf{SF}'\mathbf{F}'\boldsymbol{\phi}/2 \tag{3.9.9}$$

From (3.9.9) we note that DFs for model stability and fit, SF and AF, can be formed when the estimator $\bar{\boldsymbol{\theta}}$ is employed by adjusting $\hat{\alpha}_j$ and $\hat{\Delta}_j$ defined by (3.4.6) and (3.6.7). The specifics are left to Problem 3.21. Also, the criterion function CF, given by (3.6.2), can be reformed to meet the needs for the general estimator (see Problem 3.22).

The advantage of the general estimator $\bar{\boldsymbol{\theta}}$ can be seen if we examine the MSE of $\bar{\boldsymbol{\theta}}$. Under the true model MSE$(\bar{\boldsymbol{\theta}}) = E(\boldsymbol{\theta} - \boldsymbol{\theta})' \times$

$(\bar{\boldsymbol{\theta}} - \boldsymbol{\theta})$ can be partitioned into components as follows:

$$\text{MSE}(\bar{\boldsymbol{\theta}}) = [\text{Bias}(\bar{\boldsymbol{\theta}})]^2 + \text{Var}(\bar{\boldsymbol{\theta}}) \tag{3.9.10}$$

where

$$\text{Bias}(\bar{\boldsymbol{\theta}}) = [E(\bar{\boldsymbol{\theta}}) - \boldsymbol{\theta}]'[E(\bar{\boldsymbol{\theta}}) - \boldsymbol{\theta}] \tag{3.9.11}$$

and

$$\text{Var}(\bar{\boldsymbol{\theta}}) = E\left\{[\bar{\boldsymbol{\theta}} - E(\bar{\boldsymbol{\theta}})]'[\bar{\boldsymbol{\theta}} - E(\bar{\boldsymbol{\theta}})]\right\} \tag{3.9.12}$$

For model $f_i \in \mathcal{S}$ a proper choice of $\mathbf{D}$ results in a smaller $\text{Var}(\bar{\boldsymbol{\theta}})$, which is offset by an increase in the bias term. In examples where $\text{Var}(\bar{\boldsymbol{\theta}})$ is extremely large this trade-off is often desirable since the bias increase is commonly moderate. For this reason the estimator $\bar{\boldsymbol{\theta}}$ is referred to as a shrinkage estimator. The modeler is alerted to this situation by the parameter stability statistic DF given by (3.9.5). This decomposition is explored in Problem 3.23.

Remark 3.9.2

In the linear model setting the class of general estimators of the form (3.9.7) contains the well-known ridge regression estimator first introduced by Hoerl (1959). For $f_i \in \mathcal{S}$ taking $\mathbf{D} = \lambda\mathbf{I}$, where λ is called the shrinkage parameter, in $\bar{\boldsymbol{\theta}}$ yields the ridge estimator

$$\boldsymbol{\theta}_R = [\mathbf{x}'\mathbf{x} + \lambda\mathbf{I}]^{-1}\mathbf{x}\mathbf{Y} \tag{3.9.13}$$

Like $\bar{\boldsymbol{\theta}}$, the ridge estimator has the desirable property of shrinking the sum of the variances of the estimated coefficients at the expense of an increase in biasness. The properties and usages of $\boldsymbol{\theta}_R$ have been extensively studied since its inception and we refer to

Marquardt and Snee (1975) and Draper and Smith (1981, pp. 313-319).

Even when the fitted model function, as opposed to the estimated parameters, is the main concern of the modeler, parameter stability may be routinely investigated using PF. If instability in one or more of the parameters is found, then some adjustment is probably in order. The modeler may choose to use a shrinkage-type estimator, such as the one outlined in this section, or employ a reparametrization of the model function. In any event, a model which possesses stable parameter estimates is desirable and the DF to assess this quality, namely (3.9.5), can be used to help judge prospective models.

PROBLEMS

3.1 Meyer and Roth (1972) proposed the model $Y_j = \theta_1 + \theta_2 \exp(\theta_3 x_j)$ for the data

Y	16.7	16.8	16.9	17.1	17.2	17.4	17.6	17.9	18.1	18.7
x	1	5	10	15	20	25	30	35	40	45

(a) Use the Gauss–Newton method, (3.2.6), to find the LS estimators.

(b) Using (a), find the adjusted estimate $S^2(f) = \text{RSS}/(n - m)$.

(c) A simulation of 200 repetitions is run adding a normally distributed error with variance σ_f^2 to the fitted model $f_j = \hat{\theta}_1 + \hat{\theta}_2 \exp(\hat{\theta}_3 x_j)$. The observed dispersion matrix is the 3×3 matrix consisting of elements $\sum(\hat{\boldsymbol{\theta}}_i - \hat{\boldsymbol{\theta}})(\hat{\boldsymbol{\theta}}_i - \hat{\boldsymbol{\theta}})'/200$ where $\hat{\boldsymbol{\theta}}_i$ is the LS estimator based on the ith repetition. This matrix is

$$\begin{pmatrix} .0521 & -.0477 & 5.87 & 10^{-4} \\ & .0541 & -5.41 & 10^{-4} \\ & & 7.04 & 10^{-6} \end{pmatrix}$$

Find $\hat{\mathbf{M}}$ for the proposed model and compare $S^2(f) \times \hat{\mathbf{M}}$ to the above. Do you think $S^2(f) \times \hat{\mathbf{M}}$ is a reasonable dispersion matrix for $\hat{\boldsymbol{\theta}}$?

3.2 For modeling the S-shaped growth curves the Weibull model given by $f_j = \theta_1 - \theta_2 \exp(-\theta_3 x_j^{\theta_4})$ has been suggested. Applying this model function to the data listed in Table 3.4 results in the LS estimators $\hat{\theta}_1 = 69.96$, $\hat{\theta}_2 = 61.68$, $\hat{\theta}_3 = .0001$, and $\hat{\theta}_4 = 2.378$. At the various locations the values of $\hat{\alpha}_j$ and $\hat{\gamma}_j = \hat{\Delta}_j \times 10^7$ are as follows:

$\hat{\alpha}_j$	.3862	.3730	.3602	.3752	.4590	.3811	.3015	.2162	.1457
$\hat{\gamma}_j$	.0012	.0735	2.93	34.44	663.	2580.	2990.	2760.	1710.

Calculate the values of the DF's SF, given by (3.4.11), and AF, using (3.5.7) and ignoring $\hat{\beta}_j$, for each location where $\hat{\sigma}^2 = 1.3$. Also compute CF as defined by (3.6.2). Compare these observed DFs with those associated with the models of Example 3.4.2 (see Tables 3.6, 3.7, and 3.8). Is the Weibull model a good choice? In addition, calculate R_a^2 for the four models. Is this DF as useful as CF?

3.3 Under consistency assumptions, namely MRSS(f) – $\sum E(Y_j - f_j)^2/n \to 0$ and Var$(f)_n - \sum E(f_j - f_j)^2/n \to 0$ as

$n \to \infty$, using (3.2.13) derive an asymptotic approximation for $\mathrm{Cov}(Y - f)_n$ in terms of $\sum \Delta_j$.

3.4 For the proposed models of Example 3.4.2 calculate the nonlinearity DFs, namely NF and NF_B given by (3.7.5) and (3.7.6), respectively. Can these models be linearized at the various locations (consider Theorem 3.7.1 with $\hat{\sigma}^2 = 1.3$)?

3.5 Let $f_i \subset f_q$ and consider the test procedure of Geisser where f_q is selected if $\lambda^* < 1$ and f_i is chosen when $\lambda^* > 1$. Under hypothesized model f_i, $\lambda^* \to \lambda \exp[m(q) - m(i)]$ where λ is the likelihood ratio (3.3.3). Based on the hypotheses (3.3.1) find the asymptotic power of the test, noting as $n \to \infty$, $-2\ln(\lambda) \sim \chi^2$ with degrees of freedom $m(q) - m(i)$.

3.6 Show that the Geisser procedure is related to the Akaike (1973) information criterion (Problem 1.4 i).

3.7 Let (3.2.1) be a linear model where $f_j = \mathbf{x}_j \boldsymbol{\beta}$ where $\mathbf{x}_j$ is $1 \times m$ for $1 \leq j \leq n$.
(a) Find m, $\hat{\alpha}_j$, and $\hat{\Delta}_j$ for $j \geq 1$.
(b) Show that $\sum \hat{\alpha}_j = m$.
(c) What is the minimum value of $S_b^2(f_j)$?

3.8 Ignoring a small bias, an asymptotic confidence interval for the assumed true model value $f_j(i)$ is $\hat{f}_j(i) \pm [S_b^2(\hat{f}_j(i))]^{1/2} z_{(1-\alpha/2)}$. For Example 3.4.1 use Table 3.3 to construct approximate 95% confidence intervals for $f_j(i)$ at $x = 1$ for models f_1 and f_2. Knowing the true value is one, which model is preferred? Which confidence interval is smaller?

3.9 For a linear in parameters model f_i show that $E[\mathrm{CF}(f)] = 0$ where CF is defined in (3.6.2).

3.10 Show that if the penalty function $J(\,)$ in (3.6.1) is quadratic in $\hat{\boldsymbol{\theta}} - \boldsymbol{\theta}$ then CF is a form of $Q(i)$ given by (1.4.2). In this case find the estimator which minimizes CF.

3.11 Linearization techniques require convergence of F_j for $j \geq 1$. Estimate Var$(f_{jr}) = E(F_j - F_j)'(F_j - F_j)$ using a Taylor series approximation and (3.2.10). Construct a DF to assess this convergence.

3.12 In this problem we consider the DFs PE and IN. Based on the decomposition of Businger and Golub (1965), we write, for model f_i, $\mathbf{F}/\rho = \mathbf{QR}$ where $\mathbf{Q}$ is $n \times n$, $\rho = [\text{RSS}(f)m/(n - m]^{1/2}$, and $\mathbf{R}' = (\mathbf{R}^*, \mathbf{0})$ for upper triangular $\mathbf{R}^*_{m \times m}$. Let $\mathbf{L} = \mathbf{R}^{*-1}$ and denote the elements of $\mathbf{L}'\mathbf{h}_j\mathbf{L}/\rho$ by u_{jrs} for $1 \leq r, s \leq m$ and $j \geq 1$. With u_{jrs} we form an $n \times m \times m$ array denoted $\mathbf{U}$ and multiply $\mathbf{Q}'$ by $\mathbf{U}$ to form the $n \times m \times m$ array $\mathbf{A}$. Here the elements of $\mathbf{A}$ are $a_{irs} = \sum q_{ji} u_{jrs}$ where q_{ij} denote the elements of $\mathbf{Q}$. We partition $\mathbf{A}$ as $\mathbf{A} = (\mathbf{A}^T \mid \mathbf{A}^N)$ where $\mathbf{A}^T = (\mathbf{A}_1, \ldots, \mathbf{A}_m)$, $\mathbf{A}^N = (\mathbf{A}_{m+1}, \ldots, \mathbf{A}_n)$, and $\mathbf{A}_i$ comprises elements a_{irs}. For $\mathbf{d}_{m \times 1}$, $\|\mathbf{d}\| = 1$, let $\mathbf{V}^T_{m \times 1}$ consist of elements $\mathbf{d}'\mathbf{A}_i\mathbf{d}$ for $1 \leq i \leq m$ and $\mathbf{V}^N_{(n-m)\times 1}$ contain $d'A_id$ for $m(i) + 1 \leq i \leq n$. Here PE $= \max \|\mathbf{V}^T\|$ and IN $= \max \|\mathbf{V}^N\|$ where the maximum is taken over all unit vectors $\mathbf{d}$. An iterative method for finding IN and PE is given by Bates and Watts (1980).

(a) For the data

Observation	x	y
1	2	2.5
2	3	10.

find the LS estimator of model function $f_j = x_j^\theta$ using the Gauss–Newton method. Also, calculate ρ.

(b) For (a) the decomposition is

$$\mathbf{F}/\rho = \begin{pmatrix} -.26460 & -.96438 & -6.35 \\ .96438 & .26460 & 0.0 \end{pmatrix}$$

Find $\mathbf{A}_1$ and $\mathbf{A}_2$. How do they relate to PE and IN?

(c) Run the test of significance of PE and IN using $\alpha = .05$.

3.13 Compute CF, given by (3.6.2), for the models of Example 3.4.1 where $\text{RSS}(f) = \sigma^2[n/(n-m)]$, $\sigma^2 = .5$, and $n = 3$. Does CF detect the instability of model f_3?

3.14 Consider the nested linear model setting $f_1 \subset f_2 \subset \cdots \subset f_k$. Construct a forward selection procedure analogous to SM_B. Write theorems which take the place of Theorems 3.8.4 and 3.8.5 dealing with the probability of overfitting.

3.15 Show that $E[S^2(Q)] = \text{Var}(Q)$ where $S^2(Q)$ and $\text{Var}(Q)$ are given by (3.8.21) and (3.8.20), respectively.

3.16 In connection with Example 3.8.1 consider the choice for $\mathbf{P}(2)$ given by

$$\mathbf{P}(2) = \begin{pmatrix} .7071 & -.7071 & 0 & 0 \\ 0 & 0 & .8320 & -.5547 \end{pmatrix}$$

Show that $\mathbf{P}(2)'\mathbf{E}(2) = \mathbf{0}$. For $\beta(1) = 1$ find $E(Q)$ and $\text{Var}(Q)$, given by (3.8.20). Is this a better choice than the one utilized in the example?

3.17 Consider the LQM conditions of Borowiak discussed in Remark 3.8.2.

(a) Show that for $k = 2$ if $\mathbf{S} = \mathbf{I} - \mathbf{E}(1) - \mathbf{E}(2)$ then (3.8.26) and (3.8.30) hold.

(b) Under the choice of $\mathbf{S}$ in (a), find $\mathbf{B}(i,r)$ and show that $\mathrm{Tr}[\mathbf{B}(i,r)] = \mathrm{Tr}[\mathbf{B}(i,r)\mathbf{B}(i,r)\mathbf{B}(i,r)] = 0$. Noting condition (3.8.29), what does this imply?

3.18 In Example 3.8.4 test H_0: f_1 Correct against H_1: f_2 Correct using the exact test given by (1.5.10).

3.19 Show that the models of Example 3.8.4 are nested.

3.20 Show that $\bar{\boldsymbol{\theta}}$ defined by (3.9.7) minimizes (3.9.6) where $J(\boldsymbol{\theta}) = \boldsymbol{\theta}'\mathbf{D}\boldsymbol{\theta}$.

3.21 When the shrinkage parameter $\bar{\boldsymbol{\theta}}$ is used in place of the LS estimator $\hat{\boldsymbol{\theta}}$, using (3.9.9) find the adjusted values of $\hat{\alpha}_j$ and $\hat{\Delta}_j$.

3.22 For $f_i \in \mathcal{S}$ suppose that $\bar{\boldsymbol{\theta}}$ given by (3.9.7) is utilized. Find an asymptotic expression for $E[\mathrm{MRSS}(\hat{f}_i)]$. Can we construct a criterion function similar to (3.6.2) that, under the true model, has asymptotic expected value zero?

3.23 For $f_i \in \mathcal{S}$ let $\mathbf{P}$ be orthogonal such that $\mathbf{P}'\mathbf{F}'\mathbf{F}\mathbf{P} = \Lambda$ and $\mathbf{P}'\mathbf{D}\mathbf{P} = \gamma$ where Λ and γ are diagonal with diagonal elements λ_j and γ_j, respectively. Let $\boldsymbol{\alpha} = \mathbf{P}'\boldsymbol{\theta} = (\alpha_1, \ldots, \alpha_m)'$ and let (3.2.5) hold.

(a) Show that for large n, $\mathrm{Bias}(\bar{\boldsymbol{\theta}})$ given by (3.9.11) is $\sum \alpha_j(1 - \lambda_j/\gamma_j)$.

(b) Show that $\mathrm{Var}(\bar{\boldsymbol{\theta}})$ defined in (3.9.12), for large n, is $\sigma^2 \sum \lambda_j/[\lambda_j + \gamma_j]^2$.

(c) What is the result of large γ_j values?

(d) Show that the ridge estimator $\boldsymbol{\theta}_R$ of Remark 3.9.2 satisfies this condition for D. What is the result on $\mathrm{Var}(\boldsymbol{\theta}_R)$?

3.24 Consider the data and models of Example 3.8.4. We test (3.8.32) using the nonnested procedure of that section employing $\mathbf{P}(1)$ given in Table 3.14.

(a) Find the positive eigenvalue of $\mathbf{E}(1)\mathbf{P}(1)\Sigma_c^{-1}\mathbf{P}(1)'\mathbf{E}(1)$.

(b) Use the distribution of Q given by (3.8.17) taking $\boldsymbol{\mu} = \mathbf{0}$ to find $P(f_1 \mid f_2)$ given by (3.8.12).

(c) Use the distribution of Q given by (3.8.14) taking $\delta_i = \mathbf{S}_i'\mathbf{x}(2)\hat{\boldsymbol{\beta}}(2)$ to estimate $P(f_1 \mid f_2)$ given by (3.8.12).

(d) Compare the observed IS probabilities found in (b) and (c) and that noted in the original problem.

3.25 In this problem we outline the derivation of the rainbow test introduced in Example 3.3.2 for linear model function $\mathbf{f} = \mathbf{x}\boldsymbol{\theta}$.

(a) Define a matrix $\mathbf{D}$ such that $\mathbf{Y}^{(1)} = \mathbf{DY}$ and $\mathbf{x}^{(1)} = \mathbf{Dx}$.

(b) Consider a true model given by (3.3.7) where $W_j(\boldsymbol{\beta})$ and the vector of values is $\mathbf{W}$. If $\mathbf{Y}$ is multivariate normal with dispersion matrix $\sigma^2\mathbf{I}$ find the distribution of $\mathrm{RSS}(f \mid (\mathbf{Y}^{(1)}, \mathbf{x}^{(1)}))/\sigma^2$ and $[\mathrm{RSS}(f) - \mathrm{RSS}(f \mid (\mathbf{Y}^{(1)}, \mathbf{x}^{(1)}))]/\sigma^2$.

(c) Derive the distribution of F defined by (3.6.18).

3.26 For the data and model given in Problem 3.2 run the rainbow test outlined in Example 3.3.2. As mentioned in Section 3.6, use the five locations corresponding to the greatest LV values as defined by (3.5.9) with $\hat{\sigma}^2 = 1.3$ to form $(\mathbf{Y}^{(2)}, \mathbf{x}^{(2)})$.

4

Computer Evaluation

4.1 INTRODUCTION

Computer-aided computations have extended the horizon, in terms of new advances in modeling and model evaluation, for scientists and, in particular, statisticians. New procedures based on computer simulation and cross validation fill the recent literature, and many are based on resampling techniques. In this chapter computations made possible by the computer and simulations used in connection with model discrimination in the estimated model case are presented. These include evaluation of previously proposed DMs and DFs, most of which are based on large sample theory and normality, as well as new procedures.

Using resampling techniques, the information contained in one observed data set is extended to many typical generated data sets. With the notation introduced in Chapter 1, the modeler collects data $(\mathbf{Y}, \mathbf{x})$, where $\mathbf{Y}$ is the vector of responses and $\mathbf{x}$ is the $n \times m$ design matrix consisting of m predictor variable values at n locations. Resampling allows the modeler to construct a series of new samples which are based on the original data set. The formation of these data sets is discussed in detail in the next section. In general, the generated data sets are denoted $(\mathbf{Y}_b, \mathbf{x}_b)$, for $b = 1, \ldots, B$, and the procedure of generating these typical data sets is referred to as resampling, denoted RS.

For any proposed model from $\mathcal{S}$, given by (3.2.1), a set of statistics computed on the basis of the observed data set is collected. These include the LS estimator and the fitted model function values along with DFs which assess fit and stability. Any of these, as well as DMs themselves, are candidates for RS analysis. For each generated data set $(\mathbf{Y}_b, \mathbf{x}_b)$ statistics, which may be vector valued or functions of specific locations, denoted S_b, are computed for $1 \leq b \leq B$. The RS estimate is the mean of S_b overall B generated data sets. In addition, the dispersion matrix from RS is estimated, and its form depends on the particular RS procedure utilized. Confidence regions employing RS estimates of DFs of interest are discussed later, and for a review of RS techniques we refer to Efron and Tibshirani (1986).

Based on a hypothesized true model, RS allows analysis of relevant statistics under more than one typical data set. Properties such as convergence and asymptotic normality, common assumptions used in the construction of DFs and DMs, can be checked for any particular observed data set. This is especially useful when the DFs depend on the convergence of multidimensional variables, namely vectors and matrices. Also, the truncation of Taylor series, central to the construction of DFs in Chapter 3, can be evaluated through the accuracy of proposed DFs. When these properties are found not to hold by RS for a particular data set, new DFs based on the previously proposed methods for RS are constructed (see Geisser, 1975, and Problem 3.5).

Methods called predictive sample reuse methods and cross-validation techniques fall into the general category of RS procedures. Of particular interest because of its wide usage is the PRESS (see Myers, 1986, pp. 106–111) method for analyzing fitted models. This procedure and others are explored in the sections which follow.

Under Atkinson's question A1 or A3 the RS procedure can be used to approximate IS probabilities and hypothesis test levels. This idea is applied to the proposed DMs introduced in the previous chapter. In addition, split sample procedures, where half the data set is used for parameter estimation and the other half for model selection, are analyzed in the RS setting. Here IS probabilities are approximated.

The effectiveness of any RS procedure, whether it is used in estimation or in hypothesis testing, depends on how well the generated data sets mirror the observed data set in terms of its sampling properties. A discussion of the construction of efficient RS samples is taken up in Section 4.5. This chapter is concluded with a discussion of errors in variables models and the joint question of estimation and model selection in terms of design construction.

4.2 RESAMPLING PROCEDURES

The type of RS scheme employed by the modeler is dictated by the structure present in the original data set. Three main types of data sets, which depend on the form of the predictor variables, are considered. With each type of data set an appropriate RS method is discussed. In the sections that follow these RS techniques are utilized with DFs and DMs to evaluate prospective models. Inherent in some of these RS procedures is the application of a random number generator. A specific generator is given in Problem 4.1. Also, as in the previous chapters, the predictor variables are considered as fixed constants chosen by the modeler.

The first and least complex structure is when there exist no repeat data points, in terms of predictor variable values, in the original data set. In this case the Quenouille–Tukey jackknife method, JK, is utilized. The JK procedure consists of n repeated selections from the data set, each time deleting one point. Hence, the generated data set $(\mathbf{Y}_b, \mathbf{x}_b)$, $b = 1, \ldots, B$, takes location vectors $\mathbf{x}_b = (x_1, \ldots, x_{b-1}, x_{b+1}, \ldots, x_n)$ and response vector $\mathbf{Y}_b' = (Y_1, \ldots, Y_{b-1}, Y_{b+1}, \ldots, Y_n)$. For any proposed model from $\mathcal{S}$ the LS estimators based on the JK data sets, denoted $\hat{\boldsymbol{\theta}}_{(b)}$, are found. The statistic of interest, based on $\hat{\boldsymbol{\theta}}_{(b)}$, namely $\mathbf{S}_b$, is calculated for $1 \le b \le n$. The JK estimator uses pseudovalues for each generated data set given by $\mathbf{Q}_b = n\mathbf{S} - (n-1)\mathbf{S}_b$, or alternatively

$$\mathbf{Q}_b = \mathbf{S} + (n-1)(\mathbf{S} - \mathbf{S}_b) \tag{4.2.1}$$

where $\mathbf{S}$ is calculated on the basis of the entire data set. The JK estimator is

$$\mathbf{S}^{\mathrm{JK}} = \sum_{b=1}^{B} \mathbf{Q}_b / n \tag{4.2.2}$$

The variance of the JK estimate is

$$\mathbf{Var}(\mathbf{S}^{\mathrm{JK}}) = \sum_{b=1}^{B} (\mathbf{Q}_b - \mathbf{S}^{\mathrm{JK}})(\mathbf{Q}_b - \mathbf{S}^{\mathrm{JK}})' / [n(n-1)] \tag{4.2.3}$$

The properties of the JK estimate, such as robustness with respect to assumed distribution and homogeneity of variances, were studied by Duncan (1978) and Miller (1974), and for a full-length discussion we refer to Gray and Schucany (1972).

If, in particular, LS parameter estimation is considered, some refinement in the JK procedure is useful. As noted by Hinkley (1977), the ordinary JK estimator is unbiased but produces a larger variance, (4.2.3), than the LS estimate. This is due to the fact that the settings of the predictor variable values affect the properties of the estimator, namely the variance. For this reason weighted pseudovalues, based on the influence of the predictor variables included, have been proposed.

Following needed notation, a theorem used in the construction of these weighted values is given. Using the full data set the LS estimator $\hat{\boldsymbol{\theta}}$ is found, leading to the fitted model function values $\hat{\mathbf{f}} = \mathbf{f}(\hat{\boldsymbol{\theta}})$. Leaving out $(\mathbf{Y}_b, \mathbf{x}_b)$ produces $(n-1)\times 1$ vectors $\mathbf{Y}_{-b}, \mathbf{R}_{-b}$ and, based on $\hat{\mathbf{f}}, \hat{\mathbf{f}}_{-b}$ and $(n-1)\times m$ matrix $\hat{\mathbf{F}}_{-b}$. For the RS data set $(\mathbf{Y}_b, \mathbf{x}_b)$, deleting (Y_b, x_b), we find $\hat{\boldsymbol{\theta}}_{(b)}$ and resulting vectors, based on $\hat{\mathbf{f}}_{(b)} = \mathbf{f}(\hat{\boldsymbol{\theta}}_{(b)})$, $\hat{\mathbf{f}}_{(b)}$, and $(n-1)\times m$ matrix of first-order partial derivatives $\hat{\mathbf{F}}_{(b)}$. Also, the projected residuals using the full data set at (Y_b, x_b) is $R_{-b} = Y_{-b} - \hat{f}_{-b}$.

Theorem 4.2.1

Under the true model for large n we have the approximation

$$\hat{\theta} - \hat{\theta}_{(b)} = \hat{\mathbf{T}}_b^{-1}\hat{\mathbf{F}}_b r_b/(1-\hat{\alpha}_b) \tag{4.2.4}$$

for $1 \leq b \leq n$, where r_b is the realized value of R_b, $\hat{\alpha}_b$ is given by (3.4.6), $\hat{\mathbf{T}}_b = \hat{\mathbf{F}}'\hat{\mathbf{F}} - \hat{\mathbf{W}}_{(b)}$, and $\hat{\mathbf{W}}_{(b)}$ is the $m \times m$ matrix consisting of elements $\sum \hat{f}_{jrt} r_j$ $(r_j = y_j - \hat{f}_j)$ where the summation is over $j = 1, \ldots, b-1, b+1, \ldots, n$.

Proof For the JK generated data set (y_b, x_b) the normal equations are

$$\hat{\mathbf{F}}_{(b)}(\mathbf{Y}_{-b} - \hat{\mathbf{f}}_{(b)}) = \mathbf{0} \tag{4.2.5}$$

Expanding in Taylor series about $\boldsymbol{\theta}$ produces approximations

$$\hat{\mathbf{f}}_{(b)} = \mathbf{f}_{-b} + \hat{\mathbf{F}}_{-b}\boldsymbol{\phi}_{(b)}$$

and

$$\hat{\mathbf{F}}_{(b)} = \hat{\mathbf{F}}_{-b} + \hat{\mathbf{G}}_{-b}$$

where $\hat{\mathbf{G}}_{-b} = (\sum_t \hat{f}_{jrt}\phi_t)$, $\boldsymbol{\phi}'_b = (\phi_1, \ldots, \phi_m)$, and $\boldsymbol{\phi}_b = \hat{\boldsymbol{\theta}}_{(b)} - \hat{\boldsymbol{\theta}}$. Using these approximations and keeping only linear terms in $\boldsymbol{\phi}_{(b)}$, (4.2.5) becomes

$$\hat{\mathbf{F}}'_{-b}\mathbf{R}_{-b} + \hat{\mathbf{G}}'_{-b}\mathbf{R}_{-b} - \hat{\mathbf{F}}'_{-b}\hat{\mathbf{F}}_{-b}\boldsymbol{\phi}_b = 0$$

Noting that $\hat{\mathbf{G}}'_{-b}\mathbf{R}_{-b} = \hat{\mathbf{W}}_{-b}\boldsymbol{\phi}_b$, we find $\boldsymbol{\phi}_b = -\hat{\mathbf{J}}_b^{-1}\hat{\mathbf{F}}'_{-b}\mathbf{R}_{-b}$ where $\hat{\mathbf{J}}_b = -\hat{\mathbf{F}}'_{-b}\hat{\mathbf{F}}_{-b} + \hat{\mathbf{W}}_{-b}$. The proof is completed noting that $-\hat{\mathbf{J}}_b^{-1}\hat{\mathbf{F}}'_{-b}\mathbf{R}_{-b} = \mathbf{T}_b^{-1}\hat{\mathbf{F}}_b R_b/(1-\hat{\alpha}_b)$ (see Problem 4.2).

Utilizing the previous theorem, JK-RS can be approximated without finding the individual RS estimates $\hat{\boldsymbol{\theta}}_b$. As noted in Remark 3.2.1, $\hat{\mathbf{T}}_b$ takes the part of the observed information matrix and an alternative form of (4.2.4) substitutes $\hat{\mathbf{M}}$ for $\hat{\mathbf{T}}_b^{-1}$. Utilizing (4.2.4), the weighted pseudovalue in connection with $\hat{\boldsymbol{\theta}}$, changing $n-1$ into n, is

$$\mathbf{Q}_b = \hat{\boldsymbol{\theta}} + n\hat{\mathbf{T}}_b^{-1}\hat{\mathbf{F}}_b r_b/(1-\hat{\alpha}_b) \tag{4.2.6}$$

Here $\hat{\alpha}_b$ accounts for the leverage of $\mathbf{x}_b$ and the weighted JK estimator, denoted $\hat{\boldsymbol{\theta}}_{\mathrm{WJ}}$. Applying (4.2.2) we obtain the variance estimate $\mathrm{Var}(\hat{\boldsymbol{\theta}}_{\mathrm{WJ}})$. Theorem 4.2.1 can be applied to any statistic. For the JK data sets we let $\mathbf{S}_b = \mathbf{S}(\hat{\boldsymbol{\theta}}_{(b)})$ where $\hat{\boldsymbol{\theta}}_{(b)}$ is given by the pseudovalue (4.2.6). The JK estimator, $\mathbf{S}^{\mathrm{JK}}$, is given by (4.2.2) with $\mathbf{S}_{(b)}$ in place of $\mathbf{Q}_b$. Further, the variance of $\mathbf{S}^{\mathrm{JK}}$ is estimated by $\mathbf{S}^* \mathbf{Var}(\hat{\boldsymbol{\theta}}_{\mathrm{WJ}})\mathbf{S}^*$, where $\mathbf{S}^*$ consists of elements formed by taking the derivative of S with respect to θ_r evaluated at $\boldsymbol{\theta}_{\mathrm{WJ}}$. Inference procedures, such as confidence sets, dealing with DFs can be constructed using (4.2.6) and are discussed in the next section.

The second type of data set is where there exist repeat data points for many of the locations. As in Chapter 2, the original data set if $(Y_{qr}, \mathbf{x}_{qr})$ where r indexes repetitions, $r = 1, \ldots, n_q$, at distinct locations $q = 1, \ldots, s$. To generate RS data sets the bootstrap method (see Efron, 1979) is employed. An RS data set is generated by selecting within each distinct location q, at random, n_q data points from $\{(Y_{qr}, x_{qr}), 1 \leq q \leq n_q\}$. Thus the bootstrap procedure, denoted BS, generates data sets $(\mathbf{Y}_b, \mathbf{x}_b), 1 \leq b \leq B$, each consisting of n data points. In this way the BS data sets resemble the original data set in terms of the predictor variables' values that are present. Unlike the JK method, no weighting of pseudovalues is required.

Like the JK method, the BS procedure utilizes pseudovalues calculated on each BS sample. For statistic $\mathbf{S}$ these values are just the realizations of that statistic on (Y_b, x_b), denoted $\mathbf{S}_b^{\mathrm{BS}}$, for $1 \leq b \leq B$. The BS estimator $\mathbf{S}^{\mathrm{BS}}$ is given by (4.2.2) with $\mathbf{S}_b^{\mathrm{BS}}$ in place of $\mathbf{Q}_b$ and has variance estimate

$$\mathbf{Var}(\mathbf{S}^{\mathrm{SB}}) = \sum_{b=1}^{B} (\mathbf{S}_b^{\mathrm{BS}} - \mathbf{S}^{\mathrm{BS}})'(\mathbf{S}_b^{\mathrm{SB}} - \mathbf{S}_{\mathrm{BS}})/(b-1) \qquad (4.2.7)$$

The BS procedure has been extensively studied in the literature, and for a review we refer to Efron and Tibshirani (1986).

In general, the previous two RS procedures are robust with respect to distributional assumptions. This is not the case for the last RS method, where normality of the error and constant variance are assumed. Here the RS is done based on a hypothesized true model with known or estimated variance. For a proposed model from $\mathcal{S}$, LS estimation produces fitted model function values $\hat{f}_j = f_j(\hat{\boldsymbol{\theta}})$ for $1 \leq j \leq n$. Also, an assumed variance σ^2 or an estimate denoted $\hat{\sigma}^2$, such as from repeat data points or an estimated model function, is utilized. The model generation, MG, method forms each new data set by adding normal error with prescribed variance to the fitted model function values. Hence, $(\mathbf{Y}_b, \mathbf{x}_b)$ results from relation

$$\mathbf{Y}_b = \hat{\mathbf{f}} + \mathbf{e}_b \tag{4.2.8}$$

where $\mathbf{e}_b \sim N(0, \sigma^2\mathbf{I})$ for $1 \leq b \leq B$. The construction of the normal error terms is discussed in Problem 4.1. The MG method produces B typical data sets under the fitted model function and, as in the BS procedure, no weighting of pseudovalues is required.

With the MG method, as in the BS case, B statistics calculated from each generated data set $\mathbf{S}_b$, $b = 1, \ldots, B$, are found and play the part of the pseudovalues. The MG estimator $\mathbf{S}^{\text{MG}}$ is (4.2.2) with $\mathbf{S}_b$ substituted for $\mathbf{Q}_b$, and the variance $\mathbf{Var}(\mathbf{S}^{\text{MG}})$ takes the form of (4.2.7) ($\mathbf{S}^{\text{MG}}$ in place of $\mathbf{S}^{\text{BS}}$).

We note that the three situations and RS schemes discussed here are not the only ones available to the modeler. However, with these, the situations a modeler confronts can be handled. Another procedure is examined in the following remark.

Remark 4.2.1

Another weighted JK estimator in the linear model setting has been proposed by Wu (1986). For model (3.8.1) the LS estimator $\hat{\beta}$ is shown to take the form

$$\hat{\beta} = \Sigma|\mathbf{x}_s'\mathbf{x}_s|\hat{\beta}_s/[|\mathbf{x}'\mathbf{x}|]$$

where the summation is over all subsets s of data points (Y_j, x_j), $1 \leq j \leq n$, of size r where $r \geq m$, and $\mathbf{x}_s$ is the resulting $r \times m$ design matrix. Drawing without replacement JK samples, where $n - s$ data points are deleted, produces the weighted JK estimator

$$\hat{\beta}_{\mathrm{WJ}} = \Sigma|\mathbf{x}_s'\mathbf{x}_s|\hat{\beta}_s/\Sigma|\mathbf{x}_s'\mathbf{x}_s| \qquad (4.2.9)$$

This estimator is shown to be unbiased and the extension to smooth functions of the fitted parameters is indicated (see Problem 4.3).

In conjunction with any of these RS procedures, hypothesis tests and confidence regions for statistics of interest, such as DFs, can be constructed. Before examining specific DFs to be used in the RS model discrimination, general inference procedures are discussed. This is the topic of the next section.

4.3 RESAMPLING INFERENCE

In conjunction with the RS methods introduced in the previous section are standard statistical inference procedures. These include confidence regions, hypothesis tests, and verification of modeling assumptions. In a general format, with specific applications to the area of model discrimination, these procedures are presented in this and the following sections.

One of the basic assumptions used in the derivation of large sample statistics used in model discrimination is that of normality of the responses. This can be seen through the asymptotic properties of the LS estimator and the construction of DFs, such as SF. The fulfillment of this assumption is not a firm requirement since the discrimination statistics introduced earlier can be extended to the RS setting without normality. If, however, normality of the LS estimators or fitted model function values is a desirable property, as viewed by the modeler, it can be checked by the experimenter. This idea is developed in Problem 4.5. Also, the suggested use of the inverse of the observed information matrix $\hat{\mathbf{T}}$, given by (3.2.14), in place of the theoretical asymptotic matrix $\hat{\mathbf{M}}$, arising from Theorem 3.2.3, in DFs and DMs comes from RS evaluation (see Remark 3.2.1). This is explored through the following example.

Example 4.3.1

The compartmental data and proposed model f_2 of Example 3.3.1 are examined in terms of the form of the estimated dispersion matrix. Using the MG procedure, taking $\sigma^2 = 1.3$, a series of $B = 50$ RS data sets is constructed. The GM variance-covariance matrix, given by (4.2.7) where $\mathbf{S} = \hat{\boldsymbol{\theta}}$, and the large sample estimators $\sigma^2\hat{\mathbf{M}}^{-1}$ and $\sigma^2\hat{\mathbf{T}}^{-1}$ are found. These values are listed in Table 4.1. From these calculations we find that the observed information matrix $\hat{\mathbf{T}}$ leads to a better dispersion estimate, as it is closer to the GM-produced dispersion matrix.

For a proposed model from $\mathcal{S}$ consider statistic of interest S, where $S = s \in R$. An RS procedure produces the series of statistical values, based on each RS data set, denoted S_b for $b = 1, \ldots, B$. The simplest RS inferences are those based on the observed histogram of these RS statistics. For example, the RS $1 - \alpha$ percentile associated with S is the smallest S value, which we denote

Table 4.1 Observed and Theoretical Information Matrices

Elements	Observed inf. matrix	Theoretical inf. matrix	MG dispersion
$\mathrm{Var}(\theta_1)$	2.1229	2.9105	2.3152
$\mathrm{Cor}(\theta_1, \theta_2)$	−.0488	−.0675	−.0516
$\mathrm{Cor}(\theta_1, \theta_3)$	-3.51×10^{-3}	-4.85×10^{-3}	-3.75×10^{-3}
$\mathrm{Var}(\theta_2)$	5.77×10^{-3}	7.55×10^{-3}	5.85×10^{-3}
$\mathrm{Cor}(\theta_2, \theta_3)$	1.81×10^{-4}	2.42×10^{-4}	1.87×10^{-4}
$\mathrm{Var}(\theta_3)$	8.44×10^{-6}	1.15×10^{-5}	8.91×10^{-6}

$\mathrm{PCT}_S(1-\alpha)$, satisfying

$$\sum_{b=1}^{B} I[S_b \geq \mathrm{PCT}_S(1-\alpha)]/B \leq \alpha \tag{4.3.1}$$

where $I(A)$ is the indicator function, taking the value one if A holds and zero otherwise. Another statistic compares the observed value of S of the initial sample, denoted by the realization $S = s$, against the RS values. The RS P-value is

$$P_{\mathrm{RS}}(s) = \sum_{b=1}^{B} I(S_b > s)/B \tag{4.3.2}$$

These two RS calculations can be employed with any of the RS procedures, namely JK, BS, and MG, in which case the particular notation is substituted for RS. In a theoretical sense (4.3.1) and (4.3.2) can be viewed in the context of probability measure over the space of possible RS values.

Consider a statistic of interest S, possibly a DF. Under true model conditions, if the original sample were continuously repeated, just as with the RS-MG technique, S would possess a sampling distribution. Here it is assumed that the quantity of interest, $E(S)$, exists in terms of this theoretical RS. The construction of confidence regions on $E(S)$ based on the RS statistic S is considered. Using an RS procedure, the crudest confidence region for $E(S)$ based on large sample properties is given by

$$S^{\mathrm{RS}} \pm [\mathrm{Var}(S^{\mathrm{RS}})]^{1/2} z_{(1-\alpha/2)} \tag{4.3.3}$$

where S^{RS} is obtained from (4.2.2) and $\mathrm{Var}(S^{\mathrm{RS}})$ is its associated variance. Here it is assumed that the distribution of S^{RS} is approximately normal. Using JK and (4.2.6), another form of an asymptotic confidence interval for smooth functions of $\hat{\boldsymbol{\theta}}$ exists (see Problem 4.6). A confidence region based on the RS histogram of S relaxes this condition. Often called the percentile method, an approximate $1-\alpha$ level confidence region for $E(S)$ is

$$\mathrm{PCT}_S(1-\alpha/2) \leq E(S) \leq \mathrm{PCT}_S(\alpha/2) \tag{4.3.4}$$

where PCT is defined by (4.3.1). These two confidence regions are compared in Problem 4.7.

We note that, as is the case in nonlinear modeling, bias in estimation is generally present. For this reason other methods for interval estimation have been proposed. This is the topic of the next remark.

Remark 4.3.1

In addition to confidence regions (4.3.3) and (4.3.4), two other regions based on the histogram method have been proposed by Efron. The first, called the bias-corrected percentile method, finds $z_0 = \Phi^{-1}[1 - P_{\mathrm{RS}}(S^{\mathrm{RS}})]$ where Φ^{-1} is the inverse standard normal distribution function. The confidence region is

$$[\text{PCT}[\Phi(2z_o + z_{(1-\alpha/2)})], \qquad \text{PCT}[\Phi(2z_o + z_{(\alpha/2)})]]$$

If half of the RS values are less than S^{RS}, that is, $P_{\text{RS}}(S^{\text{RS}}) = .5$, then (4.3.5) reduces to the percentile interval (4.3.4). The other more complicated procedure is presented in Problem 4.8.

Under Atkinson's question A1 or A2 the construction of hypothesis tests utilizing RS is now considered. If the hypothesis test is based on test statistic S, the P-value of the test is simply the RS P-value given by (4.3.2). The balance of this section is devoted to examples of RS hypothesis testing in various settings. The case of two competing models is considered first.

Example 4.3.2

The compartmental modeling case introduced in Example 3.3.1 is considered. As in the original discussion, the first model is eliminated because of its poor fit and only models f_2 and f_3 are entertained. Noting that $f_2 \subset f_3$, we test hypotheses (3.3.1) taking f_2 under H_0 using the likelihood ratio test function. Using (3.3.3) we find $\lambda = 7.373 \times 10^{-5}$. Since the GLRT procedure is approximate, an RS technique is utilized to find the associated P-value. Because the data contain no repeat data points and no estimate of the variance is presented, the JK method is employed. Here the JK P-value is $P_{\text{JK}}(\lambda)$ given by (4.3.2) where $S_b = \lambda_b$ and the observed λ is 7.373×10^{-5}. This computation is left to the reader in Problem 4.9.

Other RS testing procedures are demonstrated in the next two examples.

Example 3.4.3

The case of choosing between two linear models f_i and f_r, where f_i is taken under H_0 is considered. Using an LQM from $C_{r|i}(\mathbf{P}(r))$, the conditional probability of an IS is given by (3.8.12). To avoid the complex computations inherent in (3.8.13) (using Remark 3.8.1) an RS method is utilized. We approximate the conditional probability of an IS by

$$P(f_r|f_i) = \sum_{b=1}^{B} P(\chi^2_{(1)} > q_b/4)/2$$

where q_b is the observed value of Q calculated from RS data set $(\mathbf{Y}_b, \mathbf{x}_b)$ for $b \geq 1$.

Example 3.4.4

The testing of linear models f_i and f_r under the procedure described in Theorem 3.8.3 is considered. Suppose using an LQM f_r is selected over f_i where $Q(i,r) = q(i,r) = c > 0$. The conditional probability of an IS, given by (3.8.27), can be approximated using a RS procedure. Let Δ_b be the value of $\Delta(c,r,i)$, defined by (3.8.28), based on the RS data set $(\mathbf{Y}_b, \mathbf{x}_b)$. The RS estimator is

$$P_{r|i}(c) = P[\chi^2_{(n-m(i))} \geq c/\sigma^2]\bar{\Delta} \tag{4.3.5}$$

where $\bar{\Delta} = \sum_{b=1}^{B} \Delta_b/B$. This procedure is demonstrated through Problem 4.10.

The previous RS hypothesis testing procedure can be directly extended to the multiple model setting. In the case of linear models this can be done using the framework of Example 3.8.2 or 3.8.3. When nonlinear models which cannot be linearized are contrasted, split sample procedures can be employed. This is the topic of Section 4.5. We now turn our attention to Atkinson's question A2 and consider DFs in connection with RS techniques.

4.4 RESAMPLING DISCRIMINATION FUNCTIONS

As we saw in Chapter 3, hypothesis tests for the general model discrimination problem cannot always be set up and evaluated. In the case of estimated nonlinear models question A2 is often taken up and, based on large sample theory, DFs are constructed which assess the model function's fit and stability. We now consider DFs used under Atkinson's A2 in the context of RS evaluation.

One of the most popular RS statistics for model evaluation is based on the general JK procedure. For linear model functions a predictive sample reuse statistic, called the PRESS statistic (see Myers, 1986, pp. 106-111), has been proposed. We extend this to the nonlinear model setting through JK resampling where for each generated sample $(\mathbf{Y}_b, \mathbf{x}_b)$ the pseudovalues are

$$Q_b = [Y_b - \hat{f}_{b,(b)}]^2$$

for $b = 1, \ldots, B$, where $\hat{f}_{b,(b)} = f_b(\hat{\boldsymbol{\theta}}_{(b)})$. The PRESS statistic is the JK estimator (4.2.2) times n, or

$$\text{PRESS} = \sum_{b=1}^{n}(Y_b - \hat{f}_{b,(b)})^2 \tag{4.4.1}$$

The larger the PRESS statistic, the greater the lack of fit in predicting individual deleted points by the fitted model function. This statistic is explored in Problem 4.11.

For a true model, statistic (4.4.1) can be quantified with the help of Theorem 4.2.1 in terms of its mean value. This is presented in the theorem that follows.

Theorem 4.4.1

Under the true model, for large n the approximation

$$E(\text{PRESS}) = \sigma\,[n - m + \Sigma\alpha_j(2-\alpha_j)/(1-\alpha_j)] \\ - 3\sigma^4\Sigma\Delta_j/(1-\alpha_j)^2 \qquad (4.4.2)$$

holds, where α_j and Δ_j are given by (3.4.2) and (3.4.3) using the observed information matrix.

Proof For any $b, 1 \leq b \leq B$, we expand in Taylor series about $\boldsymbol{\theta}$, keeping only linear terms in $\boldsymbol{\theta}$, to produce

$$\hat{f}_b - \hat{f}_{b,(b)} = \hat{\mathbf{F}}'_b(\hat{\boldsymbol{\theta}} - \hat{\boldsymbol{\theta}}_{(b)})$$

Using (4.2.4) we find

$$\hat{f}_b - \hat{f}_{b,(b)} = R_b[\alpha_b/(1-\alpha_b)] \qquad (4.4.3)$$

Expanding PRESS by adding and subtracting f_j and applying (4.4.3) yields

$$\text{PRESS} = \sum_{b=1}^{B} R_b^2/(1-\alpha_b)^2 \qquad (4.4.4)$$

The result is obtained using (3.5.2) and some algebraic manipulation.

The previous theorem suggests an RS statistic in connection with PRESS for nonlinear model functions. For proposed model f a DF which measures a model function's applicability to new data, based on JK, is

$$\text{PR}(\hat{f}) = \text{PRESS}/E(\text{PRESS}) \tag{4.4.5}$$

where E(PRESS) is given by (4.4.2) with the estimated statistics substituted. For efficient models this DF should be about one. As PR increases the model has difficulty in predicting future responses and is less likely to be correct. In fact, RS confidence intervals of type (4.2.4) can be constructed for (4.4.5). We also note that the form of PRESS defined by (4.4.4) leads to easy calculation (see Problem 4.11).

Example 4.4.1

An example taken from Neter and Wasserman (1974, p. 391) and explored in Myers (1986, pp. 109) concerning the sales of asphalt roof shingles and several predictor variables are considered. The data consist of 15 data points and a linear regression analysis is run. Two models examined by Myers, f_1, using predictor variables x_2 and x_3, and f_2, employing x_1, x_2, and x_3, are considered. For these models RSS, PRESS, E(PRESS), and PR are found and are listed in Table 4.2. Predictably, model f_1 with fewer independent variables has a larger RSS and PRESS than f_2. However, based on PR, defined by (4.4.5), f_1 is judged to come closer to the hypothesized true model value of one than f_2. This is due to the fact that PR accounts for the varying number of independent variables. With the help of PR the PRESS statistic is utilized. Realistically, as noted by Myers, neither model has a seriously large PRESS value, which the DF PR verifies.

The RS analog of DFs which assess the proposed model function's fit and stability is now considered. With any RS technique extensions of AF, SF, and CF introduced in Chapter 3 are formed.

Table 4.2 PRESS Statistic

Model	m	RSS	PRESS	E(PRESS)	PR
f_1	3	534.662	782.189	848.906	.921
f_2	4	272.751	643.357	529.404	1.215

From the original data set, as before, we denote the residual sum of squares by RSS($\hat{f}$) and the observed residuals by $r_j (= Y_j - \hat{f}_j)$ for $1 \leq j \leq n$. For RS data set $(\mathbf{Y}_b, \mathbf{x}_b)$ let the fitted model function's value at location $\mathbf{x}_b$ be $f_{j,b}$ for $j \geq 1$ and $1 \leq b \leq B$. The RS estimate of this value is (4.2.2), where Q_b is just $\hat{f}_{j,b}$ and is denoted $\hat{f}_j^{\mathrm{RS}}$. Also, the RSS value based on sample b is denoted RSS$(\hat{f})_b$.

The RS-DF assessing a fitted model function's accuracy at location $\mathbf{x}_j$, denoted AF^{RS}, is (4.2.2), where the pseudovalue is given by

$$Q_b = I(|Y_{j,b} - \hat{f}_{j,b}| > |r_j|) \tag{4.4.6}$$

where $Y_{j,b}$ are the responses present in the RS $(\mathbf{Y}_b, \mathbf{x}_b)$. To measure a model function's stability at location $\mathbf{x}_j$ we define SF^{RS} as (4.2.2) where the pseudovalues are

$$Q_b = (\hat{f}_{j,b} - \hat{f}_j^{\mathrm{RS}})^2 \tag{4.4.7}$$

In our last RS-DF we mirror the CF function, assessing both a proposed model's fit and stability. We define CF^{RS} as (4.2.2) where

$$Q_b = \mathrm{RSS}(\hat{f}) - \mathrm{RSS}(\hat{f})_b \tag{4.4.8}$$

Under Atkinson's question A2, applying these RS-DFs, a prospective model function can be judged relative to its competitors.

Example 4.4.2

In this example we consider the sigmoidal growth models and data of Example 3.4.2. The two stable models, namely f_1 and f_2, are analyzed in terms of the proposed RS-DFs. We note that since model f_4, introduced in Example 3.7.1, is a reparametrization of f_1 its RS statistics would be the same as those of f_1. There are no repeat data points present, but since a hypothesized σ^2 value, $\sigma^2 = 1.3$, is given the MG method of RS is utilized. For these models the MG estimators (the LS estimators averaged over all MG data sets) and the CF^{MG} values are listed in Table 4.3

We find that the MG estimators are close to the ordinary LS estimators. Also, based on CF^{MG}, model f_2 is superior to model f_1 with its value close to one. This finding is in accordance with the large sample analysis of the previous chapter. Further, at each location the RS statistics SF^{MG} and AF^{MG} are found and are listed in Table 4.4 for these models. We find that model f_2 has no serious problems in fit or stability, with only the third location having a small AF^{MG} value (.04). On the other hand, model f_1 has no stability problems with $SF^{MG} \leq .86$ at all locations but has a serious lack of fit at two locations ($AF^{MG} \simeq 0$) and marginal lack of fit ($AF^{MG} \leq .10$) at two points. This again supports the selection of f_2 over f_1 under question A2. The reader is asked to calculate the PRESS analysis in Problem 4.11.

For any particular model discrimination problem, if the asymptotic DFs presented in Chapter 3, namely SF, AF, and CF, are appli-

Table 4.3 Resample DFs

Model	Estimators			CF^{MG}
f_1	83.232	1.225	.0371	13.928
f_2	72.581	2.622	.0675	.068

Table 4.4 Resampling DFs—$SF^{MG}: AF^{MG}$

Model	Locations				
f_1	.559 : .06	.531 : .78	.459 : .2	.497 : .0	.697 : .16
f_2	.463 : .62	.474 : .56	.461 : .04	.483 : .26	.731 : .36
f_1	.535 : .16	.450 : .08	.520 : .66	.856 : 0	
f_2	.571 : .86	.459 : .26	.537 : .88	.838 : .20	

cable then the RS-DFs presented here should approximate these DFs. This is the case in Example 4.4.2. However, if the large sample DFs are not efficient because of nonnormality, nonlinearity, or slow convergence of variables, the RS-DFs proposed here provide a means of judging prospective models.

4.5 SPLIT SAMPLE PROCEDURES

If a model is selected over its rivals based on fit and stability measurements on an observed data set, its ability to explain future responses, using the same measurements, cannot be as great as on the original one. This principle is often called the optimism principle (Mosteller and Tukey, 1977). This philosophy gives rise to model function checking or cross-validation techniques. The idea of splitting the data into two parts, one for model estimation and the other for model evaluation, is not a new one (see Stone, 1974, 1978). These procedures test the ability of a fitted model function to perform on future, or new, data sets. With these procedures, tests of models under A1 or A2 can be constructed.

The major difficulty in obtaining IS probabilities for DMs involving nonnested estimated models in the dependence structure between responses and fitted model function values. In Chapter 3 this is dealt with, in the case of linearizable model functions, through conditioning on the estimated model function values and applying special conditions (see Section 3.8). The introduction of RS techniques provides the modeler with another alternative, namely split sample or two-stage procedures.

The observed data set $(\mathbf{Y}, \mathbf{x})$ is split into two groups, $(\mathbf{Y}^{(1)}, \mathbf{x}^{(1)})$ and $(\mathbf{Y}^{(2)}, \mathbf{x}^{(2)})$, consisting of n_1 and n_2 points, respectively. The first set of n_1 points is used for parameter, and hence model function, estimation, while the second set of n_2 points is used for model selection. Hence, after the model functions are estimated through $(\mathbf{Y}^{(1)}, \mathbf{x}^{(1)})$ utilizing any DM, such as an SM or LQM, a model is chosen as correct. Conditional IS probabilities are seen to be a function of the estimated model function values. Because of the splitting of the data, an independent structure between fitted model function values and the model selection process exists. Thus, RS with respect to $(\mathbf{Y}^{(1)}, \mathbf{x}^{(1)})$ can be employed, where the pseudovalues are the generated IS probabilities, to estimate conditional IS probabilities. The type of RS technique utilized, JK, BS, or MG, is dictated by the structure inherent in the original sample.

Two split sample procedures for estimating RS-IS probabilities are possible. The first, taking up A2, employs any DM and simply records the relative frequency at which particular models are selected in the RS data sets (RS with $(\mathbf{Y}^{(1)}, \mathbf{x}^{(1)})$). If one model is frequently selected, then there is enough information to make a selection. However, in this setting the quantification of the model selection is not clear since, most likely, a variety of prospective models were selected throughout the RS data sets. The second procedure is more structured, with the model selected on the basis of the original data set being fixed throughout the RS procedure. The pseudovalues are taken to be the conditional IS probabilities arising from each RS data set. This second RS evaluation is discussed for the remainder of this section.

Example 4.5.1

The case of two estimated models, f_1 and f_2, is considered. In view of Theorem 2.2.2, the LSE procedure is utilized and a model is chosen with IS probability given by (2.2.5). RS within $(\mathbf{Y}^{(1)}, \mathbf{x}^{(1)})$ produces statistics of interest S_b defined by $(\mathbf{f}(1)_b - \mathbf{f}(2)_b)'(\mathbf{f}(1)_b - \mathbf{f}(2)_b)/(2\sigma)$ for $1 \leq b \leq B$. Taking the pseudovalue Q_b as the IS probability based on S_b, the IS probability RS estimator (4.2.2) is

$$P(\mathrm{IS})^{\mathrm{RS}} = \sum_{b=1}^{B} P(Z > S_b)/B \tag{4.5.1}$$

If σ^2 is not known but an estimator exists, the RS procedure can be employed to draw this variable into the IS estimation.

The two-model case was discussed in the previous example, but extension to the multiple-model setting using simultaneous methods is straightforward. The right-hand side of (2.3.6) or (2.3.10) plays the part of the pseudovalues.

Example 4.5.2

Consider the general SM procedure where more than two models are compared. Based on the original split sample a model, say $f_{(i)}$, is selected for $1 \leq i \leq k$. Applying RS within $(\mathbf{Y}^{(1)}, \mathbf{x}^{(1)})$, the right-hand side of (2.4.14) is calculated, which we denote by $P(\mathrm{IS}(\hat{f}_{(i)}))_b$ for $b = 1, \ldots, B$. Thus, RS estimation takes $Q_b = P(\mathrm{IS}(\hat{f}_{(i)}))_b$ and yields approximation

$$P(\mathrm{IS}(\hat{f}_{(i)}))^{\mathrm{RS}} = \sum_{b=1}^{B} P(\mathrm{IS}(\hat{f}_{(i)}))_b/B \tag{4.5.2}$$

This method can be used in hypothesis testing under A1.

Example 4.5.3

This example, taken from chemical engineering (see Zabasajja, 1983), compares two models which describe the absorption of ethanol on an activated carbon surface. The Freundlich isotherm model f_1 is defined by model function $f(1) = \theta_0 X^{\theta_1}$ and the Langmuir isotherm model f_2 has model function $f(2) = \theta_0\theta_1 X/(1+\theta_1 X)$. The data are split into two parts and are listed in Table 4.5. Using the LSE method on the observed data set, model f_2 is selected over f_1. To quantify this selection in terms of the observed IS probability an RS procedure is applied. Since both X and Y are random variables and, as is typical, no repeat data points exist, the BS method of RS is utilized on the "estimated" data group. Following Example 4.5.1 using $B = 5000$ and $\sigma = 2$ (an average approximation of the variance), we find from (4.5.1) that $P(\mathrm{IS})^{\mathrm{BS}} = .111$. Thus, the Langmuir isotherm model is selected with a moderate degree of certainty.

The modeler, through RS, can estimate IS probabilities in most cases and doesn't have to weaken to A2 evaluation, namely the

Table 4.5 Ethanol Data

Estimation $x : Y$	Discrimination $x : Y$
.10: .644	.30: .821
1.18: 2.516	1.21: 2.831
1.40: 4.964	1.57: 5.232
2.87: 9.346	2.77: 9.205
4.74: 10.978	5.81: 11.327
7.10: 12.220	6.80: 12.010
11.00: 13.824	11.09: 13.973
16.30: 15.141	17.00: 15.538
19.27: 13.541	18.81: 13.251
25.30: 12.844	24.20: 13.056

construction of DFs which do not contrast model functions explicitly. But there is a price to be paid for this freedom. There is some shrinkage of these probabilities because the size of the data set used in discrimination is smaller. In addition, there is another problem in the construction of split sample designs. It is obvious that the way the data are split, in terms of n_1 and n_2, numbers of repeat data points, and specific locations sampled, affects the RS estimates through bias and conditional IS probabilities. The modeler can always test split sample designs against other single sample designs using RS Monte Carlo techniques. We follow with some general comments on the construction of efficient split sample designs.

It has been suggested in the context of model cross validation that the first part of the sample, used for estimation, should mirror the entire data set. In the case of linear models, some authors propose that the data set be split so that

$$\mathbf{x}^{(1)\prime}\mathbf{x}^{(1)}/n_1 = \mathbf{x}'\mathbf{x}/n \tag{4.5.3}$$

This implies that in connection with RS the data should be split, if possible, so that each RS data set used for model estimation mirrors the original data set as much as possible. This would reduce the bias present in estimating parameters and model function values. From (4.5.3), if the RS "estimation" locations are denoted by $\mathbf{x}_b^{(1)}$, we suggest splitting so that, as much as possible,

$$\mathbf{x}_b^{(1)\prime}\mathbf{x}_b^{(1)}/n_1 = \mathbf{x}'\mathbf{x}/n \tag{4.5.4}$$

for $1 \le b \le B$. This type of RS may be possible only in the BS or MG setting as introduced in Section 4.2. If no repeat data points exist and JK is utilized it may be possible to adjust the JK estimators using (4.2.4) (see Problem 4.12).

4.6 RELATED TOPICS

In this concluding section we touch on three topics which are commonly of interest to the modeler engaged in model discrimination. Two of these concern the violation of assumptions, or deviations from the standard setting, which we have explored so far. The third discusses the joint question of model function estimation and discrimination. These topics are complex in nature with only minimal asymptotic results known. For this reason they are best handled using RS and simulation techniques.

First we consider the violation of the assumptions imposed on the model function errors. In the derivations of the previously proposed DFs and DMs the independence and normality of the error terms were required. What if the modeler doesn't expect or doesn't wish to impose these common regression modeling assumptions? In the large sample setting, under independent errors with constant variance, NRC implies the normality of the LS estimators (see Theorem 3.2.3). This in turn results, under minimal conditions, in the asymptotic normality of the estimated model function values and justifies the DFs and DMs proposed in the previous chapters. As mentioned earlier, if the magnitude of the sample size for proper convergence is in question, the RS procedures introduced in this chapter may be employed. Hence, violations of the basic modeling assumptions can be handled through these RS procedures. If normality of the errors is in question, only the JK and BS methods are appropriate. Also, if correlations among errors are present this can be included, provide they are well defined, in the RS procedure.

The second form of nonstandard nonlinear regression modeling considered is where errors in the independent, or predictor, variables exist. The predictor variables themselves are random variables that are usually assumed to be normal with some variance, which dictates their accuracy. These models are referred to as "errors in variables" models in engineering applications and correlation models in the social, or soft, sciences. In this setting Theorems 3.2.2 and 3.2.3 do not hold, and asymptotic normality of the LS estimators under NRC cannot be assumed. An asymptotic form of the LS estimator $\hat{\boldsymbol{\theta}}$, directly showing this, is derived in Problem 4.13. If normality of the predictor variables, as well as the errors, is assumed, RS techniques, namely MG, can be successfully employed. Although the other RS methods may be utilized, we now concentrate on this RS procedure.

A model from $\mathcal{S}$ takes the modeling scheme dictated by (3.2.1) where the model function is $f_j(\boldsymbol{\theta}) = f_j(\boldsymbol{\theta}, \mathbf{X}_j)$ and $\mathbf{X}_j$ is a random vector. Normality of the predictor variables is assumed, that is, $\mathbf{X}_j \sim (\gamma_j, \Sigma_x)$. When Σ_x is known or assumed, the MG-RS procedure can be employed where the observed $\mathbf{X}_j$, denoted $\mathbf{x}_j^*$, is used in place of γ_j. The MG samples $(\mathbf{Y}_b, \mathbf{X}_b)$, $1 \le b \le n$, are formed by first obtaining generated vectors using $\mathbf{X}_b \sim N(\mathbf{X}_j^*, \Sigma_x)$. Based on these locations, the $\mathbf{Y}_b$ responses are generated using (4.2.8) where $\hat{f}_j = f_j(\mathbf{X}_b, \hat{\boldsymbol{\theta}}_b)$.

Utilizing the previously outlined MG method, the RS statistical tests and confidence regions of Section 4.3 and the RS-DFs of Section 4.4, namely $\mathrm{AF}^{\mathrm{RS}}$, $\mathrm{SF}^{\mathrm{RS}}$, and $\mathrm{CF}^{\mathrm{RS}}$, can be constructed. The effect of the errors in the predictor variables will be manifested in less conclusive results based on RS techniques. By this we mean that larger confidence regions and more conservative DFs will result. An example is outlined in Problem 4.14.

The last topic discussed is the problem of constructing efficient design, in terms of predictor variable values, for the joint problem of model function estimation and discrimination. These two goals are counter in nature. Efficient designs in terms of parameter estimation spread out the predictor variables to the edges of the design space. This results in lower leverages for the location points and smaller variances of the estimated model function values (see Kiefer, 1975). On the other hand, designs which yield small conditional IS probabilities spread the predictor variable locations evenly throughout the design space so as to pick up differences in the fitted model function values. The most noted simultaneous procedure for the selection of design points is that proposed by Box and Hill (1967), but the question remains largely an open one. For a review of these procedures we refer to the review paper of Hill (1978).

The usage of RS and simulation techniques allows the modeler to test out different design constructions, in connection with an appropriate (for the particular problem at hand) DF or DM, before the actual data are obtained. All that is required is preliminary parameter and variance estimates which can be updated when initial empirical samples are analyzed. For this purpose the MG procedure is quite helpful. Testing under any of the Atkinson's questions, the design yielding the best results can be employed.

PROBLEMS

4.1 In this problem we outline a specific random number generator. For starting seed value x_0 the algorithm is defined by relation

$$x_i = ax_{i-1}(\mathrm{mod}\,m)$$

for $i \geq 1$ where a and m are integer constants. Then $Z_i = x_i/m$ is distributed uniformly between 0 and 1. A particular choice takes $a = 2^{10} + 3$ and $m = 2^{20}$. For a full discussion of this procedure, as well as others, we refer to Carnahan, Luther, and Wilkes (1969).

(a) Write an algorithm for this random number generator. With seed value 566,387 produce a series of values. Do these look uniform?

(b) To produce a series of normal random variables with mean μ and variance σ^2 let

$$W_1 = \sigma[-2\ln(x_1)]^{1/2}\cos(2\pi x_2) + \mu$$

and

$$W_2 = \sigma[-2\ln(x_1)]^{1/2}\sin(2\pi x_2) + \mu$$

where x_1 and x_2 are uniform on [0,1]. Here W_i are $n(\mu, \sigma^2)$ (see Box and Muller, 1958). Using (a), produce a series of standard normal random variables. Do these look normal?

4.2 In the proof of Theorem 4.2.1 show $-\mathbf{J}_b^{-1}\mathbf{F}_{-b}{}'\mathbf{R}_{-b} = \mathbf{T}_b^{-1}\hat{\mathbf{F}}_b R_b/(1 - \hat{\sigma}_b)$.

4.3 Consider smooth model function, of $\boldsymbol{\theta}, f_j(\boldsymbol{\theta})$. by linearizing $f_j(\boldsymbol{\theta})$, as in Section 3.7, construct a weighted jackknife estimator of $\boldsymbol{\theta}$. What is the resulting estimator of f_j?

4.4 For the sigmoidal model function $f_j(2, \boldsymbol{\theta})$ presented in Example 3.4.2 find the JK estimator of $\boldsymbol{\theta}$, $\hat{\boldsymbol{\theta}}_{\mathrm{WJ}'}$ using the pseudovalues given by (4.2.6).

4.5 For Problem 4.4 find $\mathrm{Var}(\hat{\boldsymbol{\theta}}_{\mathrm{WJ}})$ using (4.2.3). Compare this with the large sample dispersion matrix $\hat{\sigma}^2\mathbf{T}^{-1}$ where $\hat{\sigma}^2 = 1.3$. Do you think the large sample theory holds?

4.6 Based on Problem 4.3 give an asymptotic confidence region for $f_j(\hat{\boldsymbol{\theta}})$ of the form (4.3.3).

4.7 Consider the model function and JK procedure of Problem 4.4 Construct 95% confidence intervals for $f_j(2, \boldsymbol{\theta})$ using (4.3.3) and (4.3.4) at $j = 1$ and 4. Is the large sample method, (4.3.3), accurate? Do the same with the MG procedure with $\hat{\sigma}^2 = 1.3$ and $B = 50$. Are there any differences?

4.8 An accurate confidence interval for $\boldsymbol{\theta}$ for skewed distributions is given by Efron and Tibshirani (1986). The end points of the interval are given by

$$\mathrm{PCT}[\Phi\{z_0 + (z_0 + z_{\alpha/2})/(1 - a(z_0 + z_{\alpha/2})\}]$$

where z_0 is given in Remark 4.3.1 and $a = \mathrm{Skew}(1_{\boldsymbol{\theta}}(t))/6$ and Skew is the skewness of $1_{\boldsymbol{\theta}}(t) = (\partial/\partial\boldsymbol{\theta})\log(f_j(\boldsymbol{\theta}))$ at $\boldsymbol{\theta} = \hat{\boldsymbol{\theta}}$. Find this confidence interval along with that of Remark 4.3.1 for the models and RS method of Problem 4.7 using $1 - \alpha = .95$.

4.9 Find the JK P-value using (4.3.2) for Example 4.3.2.

4.10 Consider the models and data of Example 3.8.4. Use the MG-RS method to find the IS probability outlined in Example 4.3.4. Here, take $\hat{\sigma}^2 = S^2(f_2)$. The RS estimate of $\boldsymbol{\theta}$ is put into (4.3.6) where $r = 1$ and $i = 2$ and c is the observed value of $Q(1, 2)$.

4.11 For the three sigmoidal growth models of Example 3.4.2, using (4.4.4), calculate the PRESS DF. For these models find PR(f) given by (4.4.5) with $\sigma^2 = 1.3$. Is this DF as efficient as the CF given by (3.6.2) with $\lambda = 3$?

4.12 Consider the models and data listed in Table 3.4. For models f_1 and f_2, introduced in Example 3.4.2, run a split sample DM employing the LSE criterion. Use the JK method outlined in Problem 4.4 with the two following splittings of data:

(i) $\mathbf{x}^{(1)}$ contains locations $x = 9$ through $x = 42$ while $\mathbf{x}^{(2)}$ contains predictor values 57 through 79.

(ii) $\mathbf{x}^{(1)}$ consists of values 9, 21, 57, 63, and 79. Which mirrors condition (4.5.4) more closely? Is one design better?

4.13 In this problem we derive an asymptotic form for the errors in variables model. Assume the model function is differentiable with respect to both $\boldsymbol{\theta}$ and $\mathbf{X}_j$. Let the observed value of $\mathbf{X}_j$ be $\mathbf{x}_j$ and $E(\mathbf{X}_j) = \mathbf{x}_j^*$.

(a) Expand $f_j(x_j, \hat{\boldsymbol{\theta}})$ in a Taylor series about the vectors $\boldsymbol{\theta}$ and $\mathbf{x}_j^*$ written in terms of $\boldsymbol{\phi} = \hat{\boldsymbol{\theta}} - \boldsymbol{\theta}$ and $z_j = \mathbf{x}_j - \mathbf{x}_j^*$, eliminating all but the linear terms.

(b) Using (a), rewrite the normal equations $\mathbf{F}'(\mathbf{Y} - \mathbf{f}) = \mathbf{0}$.

(c) In place of (3.2.8) of Theorem 3.2.3 impose the relation $\boldsymbol{\phi} = \mathbf{De} + \mathbf{B}$ where $\mathbf{D}$, an $m \times n$ matrix, and B, an $m \times 1$ vector, depend on $\boldsymbol{\theta}$ and $\mathbf{z}_j$ in the normal equations found in (a). Find two separate equations in terms of $\mathbf{e}$ and $\mathbf{z}_j$.

(d) From (c) find an asymptotic form of $\hat{\boldsymbol{\theta}}$ and $\boldsymbol{\phi}$. When is $\hat{\boldsymbol{\theta}}$ normal?

4.14 For the model and responses of Problem 4.14 suppose x is distributed normal with variance ^{2} where $^2 = 1.3$. Use the MG method to find values of AF^{RS}, SF^{RS}, and CF^{RS}, $\sigma^2 = .1$, .2 and .5 with $B = 50$ at each location. What effect does have in the RS statistics?

References

Akaike, H. (1970). Statistical predictor identification, *Ann. Inst. Statist. Math.*, *22*: 203–217.

Akaike, H. (1973). A new look at statistical model discrimination, *IEEE Trans. Automat. Control*, *19*: 716–723.

Allen, D. M. (1971). Mean square of prediction as a criterion for selecting variables, *Technometrics*, *13*: 469–475.

Apostel (1975). *Mathematical Analysis*, *2nd ed.*, Addison-Wesley, Reading, Massachusetts, p. 361.

Atkinson, C. A. (1969). A test for discrimination between models, *Biometrika*, *56*: 337–347.

Bailey, E. and Schmidt (1976). Estimation procedures for consecutive first order irreversible reactions, *Biometrics*, *30*: 67–75.

Bard, Y. (1974). *Nonlinear Parameter Estimation*, Academic Press, New York.

Bates, D. M., and Watts, D. G. (1980). Relative curvature measures of nonlinearity, *J. Roy. Statist. Soc. Ser. B*, *42*(1): 1–25.

Bayne, G. K., and Tan, W. Y. (1981). QOF misclassification probabilities for known population parameters, *Comm. Statist.*, *A10*(22): 2315–2326.

Borowiak, D. S., (1981). A procedure for selecting between two regression models, *Comm. Statist.*, *A10*(12): 1197–1203.

Borowiak, D.S., (1983). A multiple model discrimination procedure, *Comm. Statist.*, *A12*(24): 2911–2922.

Borowiak, D. S., and Walters, L. J. (1978). Variable input concentration models, paper presented at the Great Lakes Conference, Windsor, Canada.

Bowerman, B. L., Oconnell, R. J., and Dickey, D. A. (1986). *Linear Statistical Models, an Applied Approach*, Duxbury Press, Boston, p. 609.

Box, G. E. P., and Hill, W. J. (1967). Discrimination among mechanistic models, *Technometrics*, *9*: 57–63.

Box, G. E. P., and Muller, M. E. (1958). A note on the generation of random normal deviates, *Ann. Math. Statist. 29*: 610–611.

Box, M. J. (1971). Bias in nonlinear estimation, *J. Roy. Statist. Sec. Ser. 32*: 171–201.

Burguillo, F. J., Wright, A. J., and Burdsley, W. G. (1983). Use of the *F* test for determining the degree of enzyme-kinetic and ligand-binding data, a Monte Carlo simulation study, *Biochem. J.*, *211*: 23–34.

Carnahan, B., Luther, H.A., and Wilkes, J. (1969). *Applied Numerical Methods*, Wiley, New York.

Chambers, J. (1973). Fitting nonlinear models: Numerical techniques, *Biometrika*, *60*: 1–13.

Chien, F. W. (1981). Asymptotic theory of nonlinear least squares, *Ann. Math. Statist.*, *9*(3): 501–513.

Clarke, G. P. Y. (1980). Moments of the least squares estimators in a nonlinear regression model, *J. Roy. Statist. Soc. Ser. B*, *42*(2): 227–237.

Cook, R. D., and Tsai, C. (1985). Residuals in nonlinear regression, *Biometrika*, *72*(1): 23–29.

Cramer, H. (1946). *Mathematical Methods of Statistics*, Princeton Univ. Press, Princeton, New Jersey.

Draper, N. R., and Smith, H. (1981). *Applied Regression Analysis, 2nd ed.*, Wiley, New York.

Drygas, H. (1976). Weak and strong consistency of the least squares estimators in regression models, *Wahrsheinlich-keitstheorie und verw. Gebiete*, *34*: 119–127.

Duncan, G. T. (1978). An empirical study of jackknife-constructed confidence regions in nonlinear regression, *Technometrics*, *19*: 285–292.

Eaton, M. L. (1982). A review of selected topics in multivariate probability inequalities, *Ann. Statist.*, *10*: 11–43.

Efron, B. (1979). Bootstrap methods: Another look at jackknife, *Ann. Statist.*, *7*: 1–26.

Efron, B., and Tibshirani, R. (1986). Bootstrap methods for standard errors, confidence intervals, and other measures of statistical accuracy, *Statist. Sci.* *1*(1): 54–77.

Ehrenfeld, S. (1953). On the efficiency of experimental designs, *Ann. Math. Statist.*, *26*: 247–255.

Fox, T., Hinkley, D., and Larntz, K. (1980). Jackknifing in nonlinear regression, *Technometrics*, *22*(1): 29–33.

Geisser, S. (1975). The predictive sample reuse method with applications, *J. Am. Statist. Assoc.*, *70*: 320–328.

Good, I. J., and Gaskins, R. A. (1971). Nonparametric roughness penalties for probability densities, *Biometrika*, *58*: 255–277.

Gray and Schucany (1972)

Hill, P. D. (1978). A review of experimental design procedures for regression and model discrimination, *Technometrics*, *20*: 15–21.

Hinkley, D. V. (1977). Jackknife in unbalanced situations, *Technometrics*, *19*: 285–292.

Hoerl, A. E. (1959). Optimum solution of many variable equations, *Chem. Eng. Prog.*, *58*: 69.

Horel, G. P. (1947). On the choice of forecasting formulas, *J. Am. Statist. Assoc.* *42*: 605–611.

Imhof, J. P. (1961). Computing the distribution of quadratic forms in normal variables *Biometrika*, *48*: 417–426.

Jennrich, R. J. (1969). Asymptotic properties of nonlinear least squares estimators, *Ann. Math. Statist.*, *40*(2): 633–643.

John, J. A., and Draper, N. R. (1978). On testing for two outliers or one outlier in two-way tables, *Technometrics 20*(1): 69–78.

Kennard, R. W. (1971). A note on the C_p statistic, *Technometrics, 13*: 899–900.

Kennedy, W. J., and Gentle, J. E. (1980). *Statistical Computing*, Marcel Dekker, New York.

Kiefer, J. (1975). Optimal design: Variation in structure and performance under change of criterion, *Biometrika*, *62*(2): 277–288.

Kitagawa, G. (1979). On the use of AIC for the detection of outliers, *Technometrics*, *21*(2): 193–199.

Lai, T. L., Robbins, H., and Wei, C. Z. (1978). Strong consistency of least squares estimators in multiple regression, *Proc. Natl. Acad. Sci. USA*, *76*: 3034–3036.

Lehman, E. L. (1966). Some concepts of dependence, *Ann. Math. Statist.*, *37*: 1137–1152.

Lowere, (1982)

Malinvaud, E. (1970). *Statistical Methods of Econometrics, 2nd ed.*, North-Holland, Amsterdam.

Mallows, C. L. (1973). Some comments on C_p, *Technometrics*, *15*: 661–675.

Marquardt, D. W., and Snee, R. D. (1975). Ridge regression in practice, *Am. Statist.*, *29*: 3–19.

Meyer, R. R., and Roth, P. M. (1972). Modified damped least squares: An algorithm for non-linear estimation, *J. Inst. Math. Appl.*, *9*: 218–233.

Miller, R. G. (1974). The jackknife: A review, *Biometrika*, *61*: 1–15.

Mosteller, J., and Tukey, J. (1977). *Data Analysis Regression*, Addison-Wesley, Reading, Massachusetts.

Myers, R. H. (1986). *Classical and Modern Regression with Applications*, Duxbury Press, Boston.

Neter, J., and Wasserman, W. (1974). *Applied Linear Statistical Models*, Richard D. Irwin.

Pettersson, G., and Pettersson, I. (1970). Statistical methods for determination of empirical rate equations for enzyme reactions, *Acta Chem. Scand.*, *24*: 1275–1286.

Quenouille, M. H. (1949). Approximate tests of correlation in time-series, *J. Roy. Statist. Soc. Ser. B*, *11*, 68–84.

Ratkowsky, D. A. (1983). *Nonlinear Regression Modeling*, Marcel Dekker, New York.

Richards, E. J. (1959). A flexible growth function for empirical use, *J. Exp. Biol.*, *10*: 290–300.

Rosner, B. (1983). Percentage points for a generalized ESD-many-outlier procedure, *Technometrics*, *25*(2): 165–172.

Schwartz, G. (1978). Estimating the dimension of a model, *Ann. Statist.*, *6*: 461–464.

Searle, S. R. (1971). *Linear Models*, Wiley, New York.

Sears, F. W. (1944). *Principles of Physics*, Addison-Wesley, Reading, Massachusetts.

Shibata, S. (1976). Selection of the order of an autoregressive model by Akaike's information criterion, *Biometrika*, *63*: 117–126.

Stone, M. (1974). Cross-validatory choice and assessment of statistical predictions, *J. Roy. Statist. Soc. Ser. B*, *36*: 111–147.

Stone, M. (1977). An asymptotic equivalence of choice of model by cross-validation and Akaike's criterion, *J. Roy. Statist. Soc. Ser. B*, *39*: 44–47.

Stone, M. (1978). Cross-validation: A review, *Math. Operationsforsch. Statist.*, *9*: 127–139.

Tukey, J. W. (1958). Bias and confidence in not-quite large samples, *Ann. Math. Statist.*, *29*: 614.

Utts, J. (1982). The rainbow test for lack of fit in regression, *Comm. Statist. A*, *11*(24): 2801–2815.

Wald, A. (1943). On the efficient design of statistical investigation, *Ann. Math. Statist.*, *14*: 134–140.

Wald, A. (1949). Note on the consistency of the maximum likelihood estimator, *Ann. Math. Statist.*, *20*: 595–601.

Wilks, S. S. (1956). Sample criteria for testing equality of means, equality of variances and equality of covariates in a normal multivariate distribution, *Ann. Math. Statist.*, *17*: 257–281.

Index